Deflating Existential Consequence

Deflating Existential Consequence
A Case for Nominalism

Jody Azzouni

2004

OXFORD
UNIVERSITY PRESS

Oxford New York
Auckland Bangkok Buenos Aires Cape Town Chennai
Dar es Salaam Delhi Hong Kong Istanbul Karachi Kolkata
Kuala Lumpur Madrid Melbourne Mexico City Mumbai Nairobi
São Paulo Shanghai Taipei Tokyo Toronto

Copyright © 2004 by Oxford University Press, Inc.

Published by Oxford University Press, Inc.,
198 Madison Avenue, New York, New York 10016
www.oup.com

Oxford is a registered trademark of Oxford University Press

All rights reserved. No part of this publication
may be reproduced, stored in a retrieval system, or transmitted,
in any form or by any means,electronic, mechanical,
photocopying, recording, or otherwise, without the prior
permission of Oxford University Press.

Library of Congress Cataloging-in-Publication Data
Azzouni, Jody
Deflating existential consequence : a case for nominalism / by Jody Azzouni.
p. cm.
Includes bibliographical references and index.
ISBN 0-19-515988-8
1. Mathematics—Philosophy. 2. Nominalism. I. Title
QA9 .A94 2003
149'1—dc21   2002035831

2 4 6 8 9 7 5 3 1

Printed in the United States of America
on acid-free paper

Acknowledgments

My thanks to Jeff McConnell for urging me to take pages 49–51 of my 2000b and rework them. (He claimed the argument that appears there is too condensed to be understood. He's probably right—especially since the result turned out longer than I expected: chapters 1 and 2.) I'm also grateful to him for a number of useful suggestions regarding this material, and also for helpful suggestions that led to substantial improvements in chapter 3. My thanks to Sarah McGrath for her comments on the introduction and chapters 1 and 2, as well. My gratitude to John T. Roberts for some pointed remarks that helped sharpen an argument in chapter 2. My thanks to Mike Resnik for some off-the-cuff (but nevertheless devastating) objections to (an earlier version of) chapter 3; I hope all is well now. I'm grateful for comments from the audiences at the University of Leeds (October 17, 2001), California State University, Fresno (March 5, 2002), and the University of North Carolina at Chapel Hill (April 5, 2002), where I presented some of chapters 1 and 2; my thanks to the audience at the University of Sheffield (October 12, 2001), where I presented material from chapter 3; I'm grateful to the audience at the UCI Workshop in the Philosophy of Applied Mathematics on March 8, 2002, where I presented some of the material from chapters 8 and 9. My gratitude to my Mathematical Realism class at Tufts University in the spring of 2001, where we studied (a quite early version of) part I. And finally, my thanks to one of the anonymous reviewers at Oxford University Press for helpful comments, and to a second reviewer for helping me to see that my work (and my way of putting things) sometimes gets a rise out of people.

Contents

Introduction 3

Part I. Truth and Ontology

1 Why Empirically Indispensable Mathematical Doctrine and (Some) Scientific Law Must Be Taken as True: Preliminary Considerations 15

2 Circumventing Commitment to Truth despite Empirical Indispensability 29

3 Criteria for the Ontological Commitments of Discourse 49

4 Criteria for What Exists 81

5 Ontological Commitment and the Vernacular: Some Warnings 114

Part II. Applied Mathematics and Its Posits

6 Posits and the Epistemic Burdens They Bear 125

7 Posits and Existence 143

8 Applying Mathematics: Two Models   160
9 Applied Mathematics and Ontology   181
Conclusion   222
References   227
Index   235

Deflating Existential Consequence

# Introduction

Philosophy—at least that tradition of it I choose to work within—has *at best* an uneasy relationship with ontology. Carnap, notoriously, denied that genuine ontological questions (and purported optional answers to these questions) are meaningful, and this viewpoint is one that continues to be a live possibility (or at least a serious temptation) for contemporary philosophers. On the other hand, Quine's criterion for what a discourse is committed to is widely (and uncritically) adopted despite an official disagreement over what it amounts to, and whether a coherent version of it is even available.[1] Careful attempts to evaluate the criterion itself are sporadic and rare; conscious and unconscious reliance on it for one or another ontological project in philosophy of language, philosophy of logic, philosophy of mathematics, philosophy of science, philosophy of mind, metaphysics proper, ethics, and so on is the norm.[2]

---

[1] As a result, we find recent philosophers, e.g., van Inwagen (2000)—first writing:

> [T]here is in an important sense no such thing as Quine's criterion of ontological commitment. That is, there is no proposition, no *thesis*, that can be called "Quine's criterion of ontological commitment"—and this despite the fact that several acute and able philosophers have attempted to formulate, or to examine possible alternative formulations of, "Quine's criterion of ontological commitment" (p. 235)

—and then, without going on to tell us *how* we should understand this criterion (except to say, vaguely, that "it is a strategy or technique, not a thesis"), *using* it (or something like it?) to commit us to the existence of fictional entities.

[2] *And so on?* Absolutely: aesthetics, for example.

Of course I too understand the appeal of "getting on with it," the somber recognition that life is short, that getting to the bottom of things should not mean *staying* down there for the rest of our professional careers. It's easy to think, "let *this*, at least, be a fundamental building block in our philosophical arsenal; let *this* (at least) be something our community of philosophers can hold in common and can use as a common assumption, as a philosophical paradigm, if you will; let this (at least) unify us as a profession as we pursue our various ambitious metaphysical (or metaphysically deflating) philosophical projects."

As the last paragraph intimates, the book you're reading stays pretty near the bottom of things. Part I of this book evaluates arguments for Quine's criterion and shows them wanting. I consider various alternative criteria and eventually find a way to argue for one. The argument is detailed—tortured, even—but (I hope) ultimately convincing. I argue that Quine's attempt to substitute a logically prior criterion for what a discourse commits us to, in place of a more traditional metaphysical criterion for what exists, is a sleight of hand. Instead, I find that folk ontology takes the ontological independence of something (in a sense to be specified) as criterial for whether it exists. Accepting this requires us to overturn Quine's decision to take the objectual existential quantifier in a regimented discourse as carrying ontological commitment, and forces us to adopt a special predicate designated to carry ontological commitment instead: an *existence* predicate.

The famous Quine-Putnam indispensability thesis takes the indispensability of mathematical doctrine to scientific practice to imply (1) the truth of that doctrine, and thus to imply, via Quine's criterion for what a discourse is committed to, (2) the existence of mathematical abstracta.[3] My arguments against Quine's criterion drive a wedge between (1) and (2): the truth of mathematical doctrine (all by itself) doesn't imply the existence of the abstracta that this doctrine is about.

Traditionally, one of the best arguments *for* Platonism has been the *truth* of mathematics—either applied or unapplied. And, unsurprisingly, the standard nominalist response to mathematical truth has been either to try to show that mathematical truths aren't (or needn't be) what they appear to be (they're metaphorical, say, or they actually have very different logical forms in which obnoxious existential commitments to abstracta are missing), or to try to show that they are, if taken literally, not indispensable to empirical science (despite appearances to the contrary), and thus statements we needn't take to be *true*.

By contrast, my nominalism steals the best from both sides in this debate: I take true mathematical statements as literally *true*; I forgo attempts to show that such literally true mathematical statements are *not* indispensable to empirical science, and yet, nonetheless, I can describe

---

[3] An interesting historical point: Carnap (1939) offers what Psillos (1999, pp. 10–1) explicitly describes as an indispensability argument for "theoretical entities."

mathematical terms as referring to *nothing at all*. Without Quine's criterion to corrupt them, existential statements are innocent of ontology. (I'll temporarily call this "the separation of existential truth from ontology" or, for short, "the separation thesis.") The separation thesis results in a rather atypical nominalism, but I hope the book will show that, despite this, it's very appealing.

The main advantage of the separation thesis, overall, is that it simplifies so many metaphysical tangles. I won't illustrate this generally—there are too many topics to cover—but I do discuss (1) mathematical abstracta as they arise in the application of mathematics to science, and (2) fictional or nonexistent objects. I indicate—but hardly in a complete way—the value of the separation thesis to issues regarding *properties*. It proves equally illuminating to the metaphysics of space and time (I discuss this somewhat—but there is lots more to say, e.g., purported problems with presentism). The separation thesis provides so many simplifications in metaphysics simply because it eliminates the need to postulate something as existing just because certain *truths* prove indispensable; many metaphysical entanglements arise because this is taken for granted. Coupling the separation thesis with *other* ways of determining what exists at long last frees ontology from its linguistic straitjacket. (E.g., the linguistic form that fundamental science takes—which predicates appear in its laws—need not determine our most fundamental metaphysical commitments.)

Naturalizing Ontology

Some philosophers have been exploring what might be called *the program of naturalized ontology*,[4] the idea that in science ontological practices are already in place that dictate scientific commitments, and that to engage in "naturalized ontology" is simply to adopt *those* practices. I've *some* sympathy with this move, and broadly speaking, it's my escape route from *ontological nihilism*[5] to the safety of a particular criterion for what exists. But there are at least two crucial ways I break with this approach. First, I doubt that one finds in science *autonomous* practices exemplifying a criterion for what exists (or a criterion for what scientific discourse commits us to): Science (I claim, and I *hope*) is not as revolutionary as *that*. The criterion for what exists operating in science is drawn from epistemic and ontological practices at work in the human community at large—it's a criterion we've adopted in common. (I say a little more about this shortly.) Second, I continue to hold that the "ontological nihilism" of my 1998 is *right*: there are no philosophically conclusive ways to argue for *our* criterion for what exists. That is, we can imagine alternative communities with the same science we have but with different beliefs about

---

[4] Esp. Penelope Maddy.

[5] I borrow this depressing label for my earlier (1998) position from a (currently) unpublished paper of Penelope Maddy's.

what exists because (and solely because) they have a different criterion for what exists; they're otherwise unaffected by their choice. In particular, there is nothing we can point to, either practically speaking or in terms of some implicit incoherence in their practices or theories, that shows they've got the *wrong* criterion for what exists (see chapter 4 for a discussion of this).

Notice what's required when it's claimed that there is nothing to adjudicate between our criterion for what exists and the different one of the imagined alternative community: I'm *not* asking for an a priori resolution of the disagreement (I don't believe in the existence of things like that). I'm asking for some way of being able to show (empirically) that either we or the members of the alternative community are *worse off* because of our (or their) choice of criterion for what exists—that, for example, our (or their) scientific or epistemic practices *fall short* in some way. Only then can we say that there is a genuine issue to be judged *rationally*.

Applied Mathematics

Unlike the Quinean criterion, which commits us to everything the objectual existential quantifier in (regimented) scientific doctrine ranges over, my criterion is more discriminating. Each scientific case must be individually investigated in fair detail to see what it commits us to and what it doesn't. This is because when mathematical discourse is applied empirically, the terms in it (which indicate abstracta) can play two very different roles: sometimes they proxy for items that exist, and sometimes they don't.

As an illustration of this complex and interesting phenomenon, part II of this book examines a (small) portion of Newtonian physics. The point of this application is threefold: first, to see in a real but relatively straightforward case how my approach sorts out what exists from what doesn't; second, to see how ontological evaluation can yield principled results even in a context that far outstrips the commonsense arena that our ontological promptings were first honed in; and third, to see how powerful methodological prejudices about explanation can conspire with routine generalizing practices in mathematics to produce (when that mathematics is applied) strong (and yet unjustified) ontological intuitions. I single out the substantivalist/relationist debate about space and space-time as an excellent illustration of this last phenomenon. I'll show how the application of geometries to motion that yields simple and unified laws leads to intuitions about the existence of *absolute* space.

Principled Results

In a way, the hope for principled results, raised as my second aim earlier, brings us back to Carnap's original motivation for deflating ontology. What look like bizarre positions—at least from the commonsense point of view—are the primary offering in the history of metaphysics, as even a cursory glance at that history indicates. Monisms of various sorts, denials that humans (or any other macro-object) exist, claims that numbers exist, that *only* numbers exist, that fictional entities exist, that possible worlds exist, that *only* atoms exist, that *only* strings exist, that time doesn't exist—all such positions are available to the thrill-seeking ontologist. (Science fiction isn't half so imaginative or—according to common sense—off-the-wall as what can be found in the writings of obsessive ontologists.) Worse, and this is how I read Carnap's motivation for deflating ontology, there seems no principled way to decide among these (competing) positions.

Carnap (1950) famously distinguishes internal questions from external questions. On the latter, he writes: "This question [of the reality of the thing world itself] is raised neither by the man in the street nor by scientists, but only by philosophers. Realists give an affirmative answer, subjective idealists a negative one, and the controversy goes on for centuries without ever being solved" (p. 207).

Carnap's claim that the ordinary person (or scientist) does *not* raise ontological questions is explicitly denied by Kant (1783), who wrote:

> That the human mind would someday entirely give up metaphysical investigations is just as little to be expected, as that we would someday gladly stop all breathing so as never to take in impure air. There will therefore be metaphysics in the world at every time, and what is more, in every human being, and especially the reflective ones; metaphysics that each, in the absence of a public standard of measure, will carve out for themselves in their own manner. (p. 121)

Kant's own attempt to short-circuit un*critical* ontologizing is hardly straightforward or easy-going. But I think he was right, both about what you might call the *human instinct* for ontology and about the importance of care and subtlety in sorting out how that instinct—and ontology itself—should be treated.[6]

I didn't say so earlier, but even the Quinean take on ontology, despite its apparent rescue of that subject from the Carnapian limbo of nonsense,[7] is as deflationary and in a sense as dismissive as the Carnapian position it opposes, at least under a fairly straightforward interpretation: Ontologizing is a philosophical practice, and like all philosophical practices, it's continuous with science. Indeed, modulo regimentation, what science tells us there is, *is* what there is.

But what *does* science tell us there is? The recent outpouring of ontological posturings by professional physicists turned lay-ontologists (for examples of Quine's favorite sort of scientist) indicates nothing particu-

---

[6] I don't care much for Kant's ultimate *position*, I should add, but that's a relatively minor matter, all things considered.

[7] Quine (1951) writes that one effect of his abandoning what he describes as the two dogmas of empiricism is "a blurring of the supposed boundary between speculative metaphysics and natural science" (p. 20).

larly straightforward or definitive about ontological readings of science—at least as far as the professional practitioners of those sciences are concerned. Is the Schrödinger equation real? Weinberg thinks so.[8] Does time exist? Barbour (1999) thinks not.[9] Examples can be multiplied easily.[10] It's hard to believe that this can be sorted out neatly by regimenting scientific discourse to see what the quantifiers of the regimented discourse range over.

Why not?[11] In part because some genuine ontological concerns elude the narrower issue of what the objectual existential quantifiers in a discourse range over. This is even true of Quine's *own* ontological promptings, his physicalism, for example, which is an ontological position he takes but that *cannot*—as he admits—be represented by what objectual existential quantifiers range over in the best regimentation of the scientific discourse he accepts. To do *that* successfully, all nonphysicalistic vocabulary would have to be eliminated. And Quine admits that this can't be done.[12]

---

[8] Weinberg (1993) has a chapter he calls "Against Philosophy." Well, OK. But when he indulges in his own philosophical predilections, he offers conflicting suggestions (without, I think, quite realizing that he is doing so, or that, anyway, the whole matter requires a good deal more patience than he's willing to expend on it). About ontology he writes: "When we say that a thing is real we are simply expressing a sort of respect. We mean that the thing must be taken seriously because it can affect us in ways that are not entirely in our control and because we cannot learn about it without making an effort that goes beyond our own imagination" (p. 46). Later he attributes a view to "Scrooge," which he expresses sympathy with: "Wave functions are real for the same reasons that quarks and symmetries are—because it is useful to include them in our theories" (p. 79). (Something similar happens when he attacks social constructivist views of science. He embraces the theory-ladenness of scientific evidence-gathering practices but fails to realize that just this very concession is the primary motivation for such views.)

This is more typical than it looks. Those with already settled ontological perspectives—religious folk, for example—are usually dismissive of philosophy. *They want to get on with things.* And who can blame them?

[9] Barbour (1999) offers an ontology *Plotinus* might like. And it's striking how little concern with how ontological claims are established plays in Barbour's reasoning. Barbour thinks the main thing he has to do to show that time doesn't exist is to convince us that the Schrödinger equation governing the entire universe hasn't got an explicit variable for time. Is Ontology, in this sense, popular? You bet. Tim Folger's cover story of Barbour's views in the December 2000 issue of *Discover* is glossed on the cover of that magazine with the phrase: "Does Time Really Exist?" What follows misleadingly reads: "Why physicists secretly hope the passage of time is just an illusion." Inside (p. 61) we learn that another physicist says "cryptically" (according to Folger) that "space will go too."

Also see Belot and Earman 1999, esp. p. 178.

[10] And not just in popular books on physics. Indeed, once you know what to look for, the uncritical practice of metaphysics is ubiquitous—it even shows up in *Vogue*. (Exercise for the interested reader: Find metaphysical pronouncements in *Vogue*. Hint: Do not ignore the ads.)

[11] In pursuing this question, I'm leaving aside for the moment my settled view that the Quinean take on ontology must be rejected in any case.

[12] Quine (1978) explains why he treats the ontology of physics as special: "The answer is not that everything worth saying can be translated into the technical vocabulary of physics; not even that all good science can be translated into that vocabulary. The answer is rather this: nothing happens in the world, not the flutter of an eyelid, not the flicker of a thought,

Concerns with ontology, on the part of ordinary people, on the part of physicists trying to interpret the ontology of the wild new scientific theories they find themselves committed to the truth of, and even on the part of a grudging Quine trapped within a nonaustere web of beliefs, take the form they've always taken: queries about *which* things we *should* take to exist and which not; and this is regardless of what our discourse *seems* to commit us to.

And let's admit it: For better or worse, the ordinary person's concern with what there is and his or her concern with how we're to read off what there is from what we take to be true are real concerns just like Kant said they were, and ones to be treated with respect. However, as with the more sophisticated ontologizing that philosophers and physicists engage in, the implicit methodology that's actually doing all the work to support both sophisticated and folk ontological positions should not be left at the level of unconscious intellectual reflex or to occasional ritualized nods to philosophical ancestors, such as those to the late Quine. Rather, this methodology should be brought explicitly out into the open, where we can all participate in the evaluation of it.

One point about my *footnotes*: These are used by many philosophers solely for the purpose of citation. I don't use them that way—the reader intent on seeing the force of the position I offer should read the footnotes along with the main text. This is why I use footnotes rather than endnotes. The same is true, of course, of my previous books. But I've been told that I should point this out.

Let me conclude this introduction with a brief description of what's forthcoming. In part I, chapters 1 and 2 concern the notion of truth. My take on truth is a fairly deflationary one: The role of the truth predicate is to enable us to assent to sentences we can't explicitly exhibit. I use this role of the truth predicate plus other subsidiary considerations about scientific theories to show that applied mathematical doctrine and empirical laws must be taken by us to be true. For expository purposes, the discussion is divided into two parts. In chapter 1, I lay out the details about blind truth ascription and the truth predicate that, in a broad way, indicate why we're apt to be committed to the *truth* of what's (indispensably) contained in our body of beliefs.[13] In chapter 2, I fine-tune the issue

---

without some redistribution of microphysical states" (p. 98). Later on the same page, he describes the job of physics as "full coverage" in this sense and (still on the same page) describes his physicalism as "nonreductive" and "nontranslational." But what, we have a right to wonder, entitles *Quine* to such a position?

[13] In earlier writing, esp. my 2000b, I unthinkingly adopted Quine's phrase "web of beliefs" even as I labored to undercut the theoretical deductivism that made that phrase attractive to him. Once this terminological inconsistency was pointed out to me by Jonathan Hodge, I felt I had to find an alternative. Unfortunately, what I came up with is not as attractive as Quine's phrase.

My phrase, and Quine's too, is unfortunate in a different respect: *Belief* is a psychological term, but these beliefs are at best ones *held in common*. Alas, the naive picture of like-minded scientists peacefully assembling a system of beliefs held in common is surely wrong. The situation is complicated by various "deference practices" that detach the beliefs "held

and evaluate strategies for being agnostic or fictionalist about applied mathematical doctrine or empirical law despite its indispensability.

In chapter 3, I turn to questions of ontological commitment. Quine's criterion for what a discourse is committed to is evaluated and found wanting. There are, that is, alternatives that neither Quine nor anyone else has successfully ruled out. A substantial portion of chapter 3 is dedicated to careful evaluation of Quine's triviality thesis: the claims that "there is" is used in the vernacular to indicate ontological commitment, and that this use is regimented by the first-order existential quantifier. The conclusion is that we can't say what's the best criterion for what a discourse commits us to until we first establish a criterion for what exists.

In chapter 4, I argue that although no philosophical argument is available to definitively fix a criterion for what exists, still, it can be shown that *our* linguistic and epistemic community takes something to exist if and only if it's ontologically independent. I attempt to spell out in some detail what "ontologically independent" means.

In chapter 5, I discuss how speakers in the vernacular indicate ontological commitment. The conclusion argued for is that there are no linguistic devices, no *idioms* (not "there is," not "exists") that unequivocally indicate ontological commitment in the vernacular. Nevertheless, speakers are competent at recognizing when ontological claims are being made (provided, of course, that certain people—e.g., *philosophers*—avoid being too tricky about it).

In part II, I turn to the question of how the considerations of part I can be used to evaluate the ontological commitments of scientific theories. To this end, in chapter 6, I sort the posits (quantifier commitments) of theories into three kinds: thick, thin, and ultrathin, and distinguish these on the basis of the various epistemic requirements they impose on knowers.

In chapter 7, I argue that we take only thick and thin posits to be ontologically independent (and thus to exist). I also discuss *ontological closure conditions*, methods in science that characterize limits on what there is. I use the presence of such conditions to argue for a version of the Eleatic Principle, the claim that, in some sense, everything there is has causal powers.

In chapter 8, I describe ways in which mathematics is applied and give a number of examples, the most significant being variants on the

---

in common" from the psychological states of the individuals these "common beliefs" belong to. To sort out the sociological mechanisms involved in the detail they need goes far beyond what's possible, given the themes *this* book is supposed to cover. Nevertheless, the reader should keep in mind the implicit complexity of what's actually going on (when I refer to *our* "body of beliefs"), especially when psychological issues, about when we're committed to the truth of something or when we tentatively believe something, are brought to bear on "beliefs held in common" that aren't genuine psychological entities. I apologize for the brevity and cryptic nature of these remarks: I plan to discuss this issue more thoroughly at some future time.

I will, however, continue to use the phrase "web of beliefs" when speaking of *Quine's* views.

application of *Newtonian cohesive-body mathematics* (Ncm), a mathematical system that codifies a (small) portion of Newtonian mechanics. And finally, in chapter 9 I turn to the evaluation of the ontological status of various posits in Ncm: I argue that, for example, despite the presence of spatial and temporal posits in Ncm, such posits are ultrathin, and I also argue that Ncm isn't committed to forces, either.

I

TRUTH AND ONTOLOGY

We have to work within some conceptual scheme or other; we can switch schemes but we cannot stand apart from all of them. It is meaningless, while working within a theory, to question the reality of its objects or the truth of its laws, unless in so doing we are thinking of abandoning the theory and adopting another.

W. V. Quine

1

Why Empirically Indispensable Mathematical Doctrine and (Some) Scientific Law Must Be Taken as True: Preliminary Considerations

It's no grand matter that we take what we believe—from local statements about the cute habits of our pet hamsters to ambitious and wide-ranging theories about the workings of the cosmos—to be true. For one thing, we could change our minds about any of it tomorrow. But apart from this, a commitment to such and such a theory or statement being *true* only comes to a commitment to that very theory or statement itself.[1]

Why, then, have an idiom "is true" at all? Why talk about the "truth" of anything; why not (instead) directly *assert* what we believe whenever we need to? The answer is that sometimes talk of truth is *indispensable*: We may want to assent to a statement on pretty good grounds even though we don't know or understand what that statement is.

Terminology: An *explicit truth ascription* is any sentence of the form "'Snow is white' is true," that is, any sentence composed of the phrase "is true," preceded by a sentence within quote marks. A *blind truth ascription* is a sentence where a singular or general term—a name or description—precedes the phrase "is true." "Newton's theory of motion is true," "All the implications of the postulates of geometry are true," and "(17) is true" are examples of blind truth ascriptions. A blind truth ascription is *ineliminable* if the ascriber can't replace it with an explicit

---

[1] Some of the arguments in this chapter and the next originally appeared in my 2000b, pp. 49–51.

one by exhibiting the sentence(s) that the term in the blind ascription picks out.[2]

It's widely known that no explicit use of the truth predicate is ever indispensable. This is because *Tarski biconditionals*, the equivalences between sentences and explicit truth ascriptions to those sentences, allow us to replace explicit truth ascriptions with the sentences themselves that truth is ascribed to. From this, it follows that only blind truth ascriptions, if any, involve instances of the truth idiom that are indispensable, and this is true only of *ineliminable* blind truth ascriptions. *Are* there such?

Yes—two kinds, actually. First, consider my brilliant roommate who regularly proves mathematical truths of all sorts and who never rashly asserts anything. Suppose he steps into my room at the end of the day and tells me: "I've just proved something new." Suppose he even says what it is that he has just proved (but, of course, I don't *quite* catch it). I might say later, to someone else: "The thing Paul proved about Hilbert spaces today—whatever *those* are—is true." As long as Paul has proved one and only one thing about Hilbert spaces that day, not only does the definite description in my statement determinately refer, but such a statement is also exactly the sort of straightforward truth-attribution regularly engaged in when what's taken to be true (or false) can't itself be used explicitly. The point that it's only possible to construct an explicit truth ascription to a statement if we have "transparent access" to it (i.e., only if we're *acquainted* with the statement itself, vs. merely having a description of it) let us call the *transparency condition on explicit truth ascriptions*.

The other kind of ineliminable blind truth ascription involves assent to infinitely many sentences. If I want to assert three short sentences, I can do so by asserting a conjunction of them. If I want to assert my belief in the implications of a theory T (all infinitely many of them), I then have to say, "All the implications of T are true." Explicit truth ascriptions to infinite sets of sentences are impossible because any explicit truth ascription *must exhibit* the sentences it attributes truth to. Call this the *finite statability condition on explicit truth ascriptions*.

Of course, it's not just infinite sets of sentences that resist explicit exhibition: Finite (but large) sets of sentences can resist, as well. This is important to notice because the indispensability of the truth predicate is an indispensability to the *practice* of asserting statements. It's not right, therefore, to describe the truth predicate as dispensable in a case where someone can assent to a (long) conjunction of statements but only if given a certain outrageous amount of time (a thousand years, say). There is no principled point at which a finite set of sentences is too large to exhibit explicitly; this turns on whether simple abbreviations of them are in use, the medium in which they occur (print, orally), and so on. But

---

[2] Definitions for blind and explicit false ascriptions are similar. There is an explicit use of the truth predicate without quote marks: "It is true that. . . ." For our purposes, there's no difference between it and explicit uses with quote marks.

it's clear that, beyond a certain point, a finite set of sentences can't be exhibited explicitly, and then the truth idiom is essential to our assertion practices. Perhaps this should be called the *manageable exhibition condition on explicit truth ascriptions*. For ease of exposition in what follows, I focus only on those cases where the ineliminability of the truth idiom is due to commitments to infinite sets of sentences, and for the most part I leave aside concerns with unmanageably large finite sets of sentences. The reader should not forget them, however.

When it comes to truth ascriptions to theories or statements in *science*, it may seem obvious that it's never failure to meet the transparency condition that breeds ineliminable blind truth ascriptions; after all, scientists applying a statement or theory know *what* statement or theory they're applying.[3]

But then, why should ineliminable blind truth ascriptions ever come up in scientific practice? They can only arise—according to the foregoing reasoning—because of blind truth ascriptions to infinite sets of sentences; but aren't infinite sets of sentences, after all, rather unwieldy? No, *recursive* ones need not be: They're not if there is a perspicuous decision procedure for recognizing any member of such an infinite set. Infinite but recursive sets of sentences of this sort are (practically) quite as usable as finite sets of sentences, and in fact they come up in science *often*: The identification of a scientific or mathematical theory by a set of axioms, where that set of axioms is *infinite*, is common.

For expository purposes, I temporarily adopt an idealized description of an aspect of scientific practice, that of deducing implications from mathematized (and axiomatized) empirical theories.[4] A recursive set of axioms—that is, a set of axioms for which a decision procedure is available—can be one that's as easy to use and learn as a finite set of axioms.[5]

---

[3] Let's accept this claim for now: I show later in this chapter that some accepted scientific theories aren't explicitly exhibitable, and so blind truth ascriptions to such theories will be ineliminable.

[4] I say "mathematized empirical theory" because, so often, the language of an empirical theory involves ineliminable mathematical idioms. This deductive picture, however, is a falsification in at least two respects. First, scientific theories are often deductively *intractable*: The desired results are ones we cannot (and do not) extract from the theory by deduction. Second, scientific theories need not be well-defined axiomatic systems. I touch on both these points in what follows. See my 2000b for further details, especially on the first observation.

[5] Notoriously, Davidson 1965 seems to claim otherwise. *Pace* that view, the axioms for Peano arithmetic (PA), e.g., include an *induction schema*: $[A(0) \& (x)(A(x) \to A(x'))] \to (x)A(x)$, an infinite set of axioms that arise from this schema by the substitution of any formula (with the variable x free) for $A(x)$. This infinite set of sentences is particularly perspicuous because it contains only substitution-instances of a schema. A less perspicuous set of sentences (but equally usable and routinely studied) is the set of tautologies of any first-order language. Axiomatizations of such languages usually just include tautologies wholesale (because truth tables provide an easy decision procedure for them). See, e.g., Enderton 1972.

Useful terminology. A set of sentences T is *finitely axiomatizable* if there is a finite set of sentences from which the sentences of T can be deduced. PA is not finitely axiomatizable.

This is because in both the finite and infinitely recursive cases, mechanical methods are available for verifying that purported proofs are valid and sound.[6]

But, and this is the crucial point, although implications can be drawn from PA (say) just as easily as from a finitely axiomatized theory T, blind truth ascriptions can be eliminable in the second case, although not in the first.

For consider: Presume a commitment to the truth of T, and thus to the truth of some implication of T, which I represent here by the schematic symbol "*Im*." Provided T has a manageably exhibitable conjunction of its axioms, which I represent by the schematic letter "A," the informal description of our truth-commitments just given (involving "truth") can be eliminated by using the sentences themselves instead: A. $(A \rightarrow Im)$. *Im*.[7] In the first case, however (where "*Im*" represents an implication of PA), we *must* say

PA *is true*. PA implies '*Im*'. '*Im*' is true. *Im*.

(The last step requires the *Tarski biconditional*: '*Im*' is true $\Leftrightarrow$ *Im*.) Because PA is not finitely axiomatizable, it can't be exhibited explicitly; the use of the truth idiom in this inference is ineliminable.

A study of the axiomatizability of first-order theories reveals how quickly and easily recursively but not finitely axiomatizable mathematical theories come up.[8] Whenever such theories are used to draw implications, blind truth ascriptions are ineliminable.

The use of schemas where a finite axiomatization isn't available is not restricted to applied mathematical doctrine, however. Consider Newton's

---

[6] Because not all mechanical methods are equally practical, an infinite set of sentences can sometimes be more useful than a quite large finite set of sentences. The former may come in an especially perspicuous form (e.g., iterated conjunctions of the single sentence "Peter is running") that makes the decision procedure for them easy, whereas a list for the latter, even though finite, may be quite difficult to peruse.

[7] To do this we identify T with a particular conjunction of its set of axioms. In general T can be (finitely) axiomatized, and the axioms ordered, in more than one way. Similarly, a statement can be expressed in more than one way. In both cases, fixation on a particular incarnation of a statement or theory for purposes of assertion is harmless.

In what's forthcoming, "'*Im*'" is a schematic symbol representing the Tarskian metalanguage name of the implication that the schematic symbol "*Im*" represents. "PA" will function as a name for Peano Arithmetic both in the discourse of this book and in the discourse containing the inferences I give as examples. This will also be true of variables standing for theories that arise later in this chapter.

I should say that I won't be particularly fussy about use/mention matters, quotation, or quasi-quotation except when I fear misunderstanding.

[8] See, e.g., Chang and Keisler 1973, pp. 41–3. Although PA is an appealing illustration because it's an intuitively simple axiomatization of ordinary arithmetic, and yet is both incomplete and not finitely axiomatizable, there is, in general, no connection between finite axiomatizability and completeness. Q (p. 43) is a well-known subtheory of PA that is finitely axiomatizable and incomplete. Presburger arithmetic, although not finitely axiomatizable, is complete.

second law, $\mathbf{F} = m\mathbf{a}$. This, as many have noted, seems to be a schema. Kuhn (1970) writes: "That expression proves on examination to be a law-sketch or a law-schema" (p. 188) and he points out (pp. 188–9) that in the case of free fall, the expression becomes $mg = m(d^2s/dt^2)$; in the case of the simple pendulum, it's transformed to $mg\sin\theta = ml(d^2\theta/dt^2)$; and so on. This proves to be true of many laws: They're law-schemas, which when instantiated—often jointly—become statements we're committed to.[9]

Is it reasonable to describe (this piece of) Newtonian physics as not finitely axiomatizable? After all, the mere use of schemas (instead of a finite set of axioms) to codify statements of a theory doesn't by itself show that the theory is not finitely axiomatizable. One must show that *no* finite axiomatization exists—often a nontrivial matter. It's relatively easy to see, however, that instantiations of Newton's second law can't be finitely axiomatized.

Consider the set of idealized situations where the only force law is universal gravitation: $\mathbf{F} = -(Gm_1m_2/r^2)\check{\mathbf{r}}$, where G is the gravitational constant, m_1 and m_2 are the masses of two point-masses M_1 and M_2, r is the distance between M_1 and M_2, and $\check{\mathbf{r}}$ is the unit vector from M_1 toward M_2 and where the only things are point-masses of varying masses, varying initial velocities, and varying initial distances from each other. Notice that we can instantiate Newton's laws (resulting in a set of differential equations) for each one of these situations, but that situations that differ in their numbers of point-masses are *logically* independent of each other. If the theory were finitely axiomatizable, it would be possible to (logically) deduce the differential equations governing situations containing any number of point-masses from a finite set of characterizations of situations containing specific numbers of point-masses at specific locations, with specific masses, and with specific initial velocities.

The foregoing argument for the ineliminability of the truth idiom in scientific practice turns on the use of recursively infinite sets of sentences, either in the mathematics applied or in the physical theory itself. But these examples of infinite sets of sentences are ones codifiable in *schemas*. Schemas certainly *are* finitary ways of coding infinite sets of sentences. Why, then, can't one express commitment to an infinite set of sentences by asserting the *schema* representing them (and so, in this way, bypass the need for the truth idiom)?

---

[9] A substantial discussion of schemas in scientific contexts may be found in Koslow 2000, although my understanding of "schema" differs from his. He describes Newton's second law as schematic because it leaves open what sorts of forces there can be. I agree but regard an instantiation of any law as still schematic unless *all* the parameters in it have values. (One of Kuhn's examples, as well, is not an instantiation in my sense; what's missing is the substitution of a specific scalar for "*l*.") Koslow also takes schemas to have meanings, whereas I adopt the Quine line that—strictly speaking—schemas don't *say* anything themselves but *stand for* a set of sentences. See what's forthcoming later in this chapter.

Arnold Koslow tells me in an e-mail dated October 2000 that Sidney Morgenbesser had made the point (to him) about the schematic character of physical laws as early as 1955.

Quine has claimed again and again that schemas themselves can't be used to express *anything*.[10] His point is that schemas are only (diagrammatic) tools used to talk *about* the sentences of a language: they allow us to codify an infinite set of sentences and to talk about that set in the context of a metalanguage. If, therefore, we're interested in characterizing a set of sentences by schemas and then assenting to them, the truth idiom is still needed, and what results is a blind ascription (e.g., "All the instances of the induction schema are true").

One recalls how notoriously finicky Quine is about use/mention errors; he's positively frightened that someone will unconsciously treat schematic letters as variables, bind them with quantifiers, and in this way inadvertently adopt a higher-order theory of some sort. So it's true that in *Quine's* hands schemas—with their "dummy letters"—are understood as diagrams that—horrors!—one had better not quantify into unless one wants to slip into unintended commitments to attributes (or worse). Consequently, on this view, schemas are not statements and can't be used to assert collections of sentences. But why share Quine's terminological puritanism?

Instead, let's introduce *substitutional quantifiers* (Σ and Π, say) with substitution-classes containing the sentences of the language, predicates holding of classes of those sentences, and special quotation devices which the substitutional quantifiers can bind variables within. If "I" is impounded as a predicate so that all substitutions of instances of the induction schema of PA for "x" in "$I'x'$" yield truths, we can then commit ourselves to all the instances of the induction schema by asserting: $(\Pi p)(I'p' \to p)$.[11]

Unfortunately, if the aim is to circumvent indispensability of the *truth idiom*, this fails because once substitutional quantifiers are available, we can define truth outright.[12] Moral: Dressing up the truth idiom in imposing notation won't dispense with it.

A similar point applies to those tempted to reach out and grab what they need: The truth idiom is needed to assert a recursively infinite collec-

---

[10] Quine 1941, pp. 18–9; 1986, p. 67; 1955, pp. 153–4; 1972a; and esp. 1972b, pp. 272–3. In Quine (1980) he says of sentential calculus schemas: "No meaning is to be attached to such expressions: they serve only in the manner of diagrams, in connection with general discussions of truth-functional structure" (p. 36). Similar passages occur often in his corpus.

[11] *Useful terminology. Objectual quantifiers* come with a given domain of objects that they range over. An existential sentence with an objectual quantifier (e.g., $(\exists x)Sx$) is true (roughly) if there is something in the domain that is in the extension of the formula Sx. *Substitutional quantifiers* come with a substitution class of terms that can be substituted for the substitutional variable. An existential sentence with a substitutional quantifier (e.g., $(\Sigma X)SX$) is true (roughly) if there is a term O in the substitution class of the quantifier and if the sentence SO is true. As Quine (1969, p. 106) points out, any set of terms can serve as a substitution class, even the left parenthesis.

[12] Given equality, we can define truth thus: $Ta \Leftrightarrow (\Sigma p)(p \ \& \ 'p' = a)$, where a is a name-schema of sentences in the substitution class of (Σp).

tion of sentences? Fine, let's just assert the collection of sentences itself by means of an *infinite conjunction* of those sentences. Perhaps it can even be argued that this is routine in ordinary language: We indicate how additional cases can be constructed, and then we wave our hands and say, "and so on." But the infinitary conjunction also gives us a logical idiom from which a truth predicate can be directly defined—just as before. In both cases, we don't escape the truth predicate as much as express it by other means.

One last strategy of this type may have occurred to the reader. Finite axiomatizability seems relative to the *ontology* (in the Quinean sense) of the theory in question. Introduce more *objects* of a certain sort (quantification over *sets* or *properties*, in the case of PA; quantification over *forces*, in the case of (the bit of) Newtonian mechanics given above; quantification over *propositions* or *sentences* in the case of tautologies of the sentential calculus), and then finite axiomatizability (of the resulting theory) follows. *Logically* speaking, that is, indispensable use of the truth predicate isn't very deep: Essential use of the truth predicate can be eliminated by introducing more substantial quantifier commitments. A quite similar strategy is to argue that the finite-axiomatizability problem is a parochial one due to the choice of first-order logic. Use a higher-order logic instead; second-order PA, for example, is finitely axiomatizable.[13]

One argument that seems available against this (and which I'm sympathetic to) is that, again, the truth predicate has not really been eliminated; rather, its powers have been tucked into another notation, into the *ontology*, as it were. This is clearest and most explicit in the case of the second-order move: The second-order version of PA is more powerful than first-order PA; that it's more powerful is provable because the truth predicate for first-order PA is definable in second-order PA.[14]

As I said, I'd like to apply this point to cases where the language remains first order but the ontology (in Quine's sense) has been augmented. This may not be convincing, however, in all such cases: The postulation of propositions or forces can be motivated apart from the

---

[13] The induction schema is replaced by the second-order sentence: $\forall X[(X0 \;\&\; \forall x(Xx \to Xsx)) \to \forall xXx]$.

[14] See, e.g., Shapiro 1991, p. 110. The moral here, and with respect to the previous approaches just considered, is that logical devices that enable us to bypass the truth idiom are (pretty much) powerful enough to define that idiom in their own terms. This shouldn't be surprising: There's a sense in which the truth predicate is the minimally powerful logical device for the purposes it serves.

One caveat: Of course second-order PA provides (at least directly) neither a truth predicate for itself nor a truth predicate for the (empirical) sentences to which number theory is to be applied. On the first point, this is the same situation that we find with (predicative) truth idioms generally. The second point can, in general, usually be finessed by extending the range of the truth predicate (e.g., for first-order PA) to include those (true) sentences. As discussed in part II, this is, in any case, relatively unimportant because most applications of mathematics to empirical subject matters involve, in a very real sense, *absorbing* those empirical subject matters into that of the mathematics being applied.

desire to avoid ineliminable blind truth ascriptions. And so, rather than attempt the uphill struggle of arguing that quantification over a type of thing amounts (in certain cases) to an implicit coding of a truth predicate, I'd prefer to go a different route. This is, first, to show that there are numerous cases in empirical science where indispensable blind truth ascriptions come up in a way that can't be eliminated by the supplementation of ontology in Quine's sense. I do this in the very next paragraph. My second move is to show that if the role of the truth predicate is for the purposes of blind truth ascription, then even if indispensable use of the truth predicate (in certain cases) is absent, this doesn't absolve us of a commitment to *truths*. This argument is given in chapter 2 (see esp. the summary and concluding remarks).

I've allowed myself (up until now) a description of scientific theories where their implications are recognized via deduction. But as footnote 4 indicated, science is more complicated; we must often use tools other than deduction to find the implications of a theory. I now discuss certain aspects of this that reveal how the transparency condition contributes to the indispensability of the truth idiom to scientific practice.

There is a tendency to divide scientific methodology discretely into induction and deduction. Crudely characterized, the view takes the deductive side of scientific methodology to be one where a theory is in place and implications are deduced from it (using logic). Of course, background mathematics is in place, as well, but this is neatly described—on the deductive picture—as just so many more premises in deductions of the consequences of a scientific theory.

Unfortunately, deductive consequences about a type of situation are often out of reach: Using the above talk of instantiations of physical-law-schemas, either (a) the instantiation (of the law-schemas in question) *can* be written down, but the mathematics involved makes the instantiation too intractable to extract the desired consequences from,[15] or (b) an accurate description of the physical situation is too complicated to allow us to write down the instance to begin with (the required parameters and constraints are too complicated and numerous), or worse, we're simply ignorant of crucial details, and so we're prevented from giving the instance (the actual shapes of the molecules are too irregular to discover the details of, the actual intermolecular forces are too complicated to get a grip on). In case (b), notice, the desired instantiation *itself* can't be

---

[15] Sussman and Wisdom (1992):

> Our physical model is the same as that of QTD except in our treatment of the effects of general relativity. General relativistic corrections can be written in Hamiltonian form, but we have not been able to integrate them analytically. Instead we used the potential approximation of Nobili and Roxburgh ... which is easily integrated, but only approximates the relativistic corrections to the secular evolution of the shape and orientation of the orbit. (p. 57)

written down: Although codifications (via schemas) of the physical laws *can* be written down, not all the instances of those schemas can be.[16]

The glory of science (*one* glory, anyway) is that in both these cases, we're *not* prevented from applying scientific law: Other tools enable us either (a) to justify the use of formulas that are—strictly speaking—*not* implications of the theory (such formulas are what may be called "approximation-law-schemas") or (b) to recognize that certain implications exist despite an inability to deduce them from the instantiation. I illustrate both these practices briefly.[17]

(1) *Calculational shortcuts can be used to derive formulas (approximation-law-schemas) that are known to be only "approximately" true of the situations they're applied to.* Often, in physics, formulas are derived by the substitution of terms for other terms (e.g., the substitution of θ for $\sin\theta$ when θ is "small") or by the simple but outright elimination of some of the mathematical complexities in formulas (e.g., dropping everything but the first few terms in an infinite Taylor series expansion). The resulting mathematically neat formulas are easily applied in a broad range of situations; the results they predict deviate slightly enough from empirically derived results that differences can be ignored.

For example, consider two concentric circular loops of wire, the outer with a slowly changing current that generates a magnetic field, and the inner in which a current is induced. A formula for how the electromagnetic force in the inner loop depends on the changing current in the outer loop is easily derived, provided that the current is changing "slowly enough" and that the inner radius of the loop is "small" compared with the radius of the outer loop. Why? Because to derive the formula, it's assumed that the currents are steady ones (i.e., in practice, the application of the formula to a situation where the current is changing—provided it's changing slowly enough—will induce only slight discrepancies between empirical results and what the formula predicts), and because the variation of the field in the interior of smaller loop is also ignored in deriving the formula (and again, if the inner loop is *small enough* compared with the outer loop, then only slight deviations from what the resulting formula predicts will result when the formula is applied in such cases).[18]

(2) *Reasoning about simplified models of a theory-schema can be used to show that an instantiation of that theory-schema has certain implications*

---

[16] This should be familiar. For all practical purposes—and practical purposes are what matter when asserting statements—most instances of the induction schema can't be written down either. (They're too long.)

[17] These examples are drawn and modified from my 2000b, pt. 1, § 2. It's worth mentioning that these tricks—and others—are often applied simultaneously.

[18] For details on the derivation, see Purcell (1985, pp. 276–9). Other examples, such as Ohm's law and self-inductance, are in Purcell (1985, pp. 128–33, 283). Also see Feynman et al. (1963, 15–10) for approximating formulas in relativistic dynamics, oscillations in various contexts (23–4, 25–5, 30–11), polarization of light (33–9), etc. All these exam-

even if they can't be explicitly drawn. Instead of fruitlessly trying to deduce consequences from an instantiation that's mathematically intractable or, worse, can't even be written down, we instead use an instantiation of the theory-schema with respect to some other (simplified) situation (actual shapes of molecules are replaced with spherical ones; vaguely elliptically shaped lumps of inhomogeneous magnetic material are replaced by exactly elliptically shaped homogeneous lumps) and draw implications from that instantiation instead.[19] Subsidiary arguments then show that because certain implications follow from the written instantiation, it's reasonable to think that those implications—or suitably modified implications—follow from the unwritten instantiation, as well. Implications of a theory, that is, are recognized by heuristic arguments involving the existence of *other* implications of the theory, not by drawing *the* implications actually wanted.

One way to argue that the desired implications of a theory actually exist is to provide a *sequence* of models corresponding to different instantiations, where later ones are refinements of earlier ones in the sense that some of the real-world complications represented in the mathematical formalism of later models are aspects missing from earlier models—and then one shows that (some of) the deviations between prediction and empirically derived result, which one takes to be due to idealizations in the earlier model, have indeed disappeared.

What do these observations indicate about the ineliminability of the truth idiom? I claim that in both cases physicists engage in reasoning *about* their theories as much as they reason *within* them. With (1), the idea is that certain formulas will give results close to *what's true* because the shortcuts and modifications used to construct the neat formulas are ones that involve aspects of the phenomena that fall within certain (desirable) measurement thresholds. But, often, the informal phrase "what's true" can't be replaced with anything explicit because the point of introducing the neat (false) formulas is that the exact right formulas are often unavailable (underivable).

One may think that references to "what's true" in these cases are eliminable in terms of an explicit requirement of a (numerical) match between empirically recognized results (what we discover by observation,

---

ples have the general form just described: An aspect of the phenomenon is ignored, and if its numerical contribution to the resulting effect is sufficiently tiny, then using the derived formula will result in deviations small enough to live with (given our purposes, and measurement capacities, at that time).

[19] Often, not merely is the instantiation simpler, it may be—strictly speaking—an instantiation of *a simpler set of law-schemas*. For example, we may neglect a certain type of force altogether (e.g., in calculating the trajectory of certain planets, we not only leave out a number of other small bodies—which we presume won't have much impact—but also ignore the effects of electromagnetic forces, because those are also small). In effect, we replace a model of a world where certain laws apply with one in which those laws don't apply at all.

measurement, instruments, etc.) and what the formulas predict. Indeed, my discussion of (1) may seem to indicate that this is the right way to describe things: We want neat formulas that agree with empirically recognized results (to within certain thresholds). Why then drag in talk of *truth*?

The problem with this way of putting the matter is that *empirically recognized results* are themselves not the final court of appeal, because such results are open to distortions due to incorrectly functioning instruments, human error, and so forth. The point isn't to get neat formulas to agree with empirically recognized results (although that's how we proceed in practice); the point is to get neat formulas to agree with empirically recognized results that are *right* (to within certain thresholds of error). This talk of *what's right*, however, is (barely disguised) talk of truth *itself*.

The second case may look different. One imagines it this way: (a) We have *in hand* an instantiation A of (the) physical laws governing a particular simplified situation, (b) we *know* that an instantiation B of (the) physical laws governing a particular complicated situation *exists*, (c) we have *in hand* an actual derivation of C from A, and (d) as a result, we can mount a (complex) argument that therefore a claim C' (related to C in certain respects) follows from B. Why should a blind truth ascription come up here? Well, it needn't if we can write B down. But, as I've mentioned, the problem in practice is often (although not *always*) that we can't write B down. In these cases, the advantage of A is that it *can* be written down (and so there's a hope of directly drawing implications from it, in contrast to B).[20]

This brings us to a point worth making explicit. As long as the idealized view of scientific practice was in view, it looked like ineliminable blind truth ascriptions are only required (in scientific practice) because of the need to express a commitment to theories that can't be finitely axiomatized (or are unmanageable). But as soon as it's recognized that we often draw inferences from instantiations of theories that *can't be written down*, it's clear that we're forced to accept blind truth ascriptions simply because we're ignorant of what the instantiation (we're studying) actually looks like. And the (very) interesting point is that lots of physics involves discovering implications of instantiations of physical laws even though those instantiations can't be exhibited in explicit acts of deduction. Notice that, because of this, the strategy of trading in ineliminable blind truth ascriptions for an augmented ontology (in the Quinean sense) isn't available in

---

[20] Sheer geometry can pose problems for the instantiation of law-schemas: E.g., contours are shaped in all sorts of ways—but the number of contours that can be neatly and precisely characterized by tractable mathematical equations is tiny by comparison. Often, however, a mathematical description of the shape of a contour must be inserted into a law-schema in order to construct an instance (e.g., if a current is traveling along a wire bordering an area). So, in practice, contours that can't be (tractably) described by mathematical equations must be replaced by ones that can be.

these cases. Such a strategy can't be carried out unless the theory can *at least* be indicated in schematic form.

Something even odder is possible. We may be unable to write down the physical-law-schemas *themselves* even though we're able to recognize instances of those schemas (in certain idealized situations) or at least implications of instances of those law-schemas. Can we become *committed* to a scientific theory in a case where we can't write down a physical-law-schema for it? Of course. We may find that all the implications we're able to recognize as following from instances of such a theory-schema are ones that convince us that all the instances of that theory-schema are true. In such a case, commitment to the truth of the theory can follow despite the inexpressibility of the theory—even in schematic form.

Some General Remarks about "True"

In chapter 2, I consider a large number of additional strategies aimed at circumventing any argument (to the effect that we're committed to the truth of indispensable scientific law and empirically applied mathematical doctrine) that is based on the necessity of blind truth ascriptions. But what I want to do now is to make explicit a few observations about the truth predicate that follow from the mere requirement that it enable blind truth ascriptions.

The first point to make right at the start is that in claiming that the truth predicate enables blind ascriptions, I'm *not* claiming that the truth predicate manages this feat all on its own. As ordinary examples of blind truth ascription make clear, the truth predicate enables blind ascription in the vernacular in coordination with other devices, such as quantifiers ("Everything John says is true"), names ("(17) is false"), and various kinds of indicators ("That sentence on the chalkboard is false"). Strictly speaking, the role of the truth predicate is that of facilitating semantic ascent and descent. That is, to use language Quine was so fond of, it allows us to move smoothly between use and mention.

It's this that makes the Tarski biconditionals seem so central to the meaning of the truth predicate. But rather than take the Tarski biconditionals as "constitutive" of the meaning of the truth predicate or as "a priori" constraints on that predicate, it's better to see, when blind truth ascriptions are made to sentences *of our own language*, how the Tarski biconditionals are essential to the task.[21] John, let us presume, is capable of uttering any sentence of our language. And let us also presume that

---

[21] Why the *explicit* restriction of the examples to *our own* language? First, because "true" as used in the vernacular applies to sentences whether they occur in our language or not; the value of blind truth ascription isn't exhausted by its application to our own language. But, second, what licenses the truth predicate in its application to other languages can't be those Tarski biconditionals expressible in our own language. So there is a serious sense in which the logic of the truth predicate (in the vernacular) is *not* captured by Tarski biconditionals. See my 2001 for details about this.

Bob regards John as a totally sober and trustworthy individual. Then, for any sentence *A* of our language, Bob can engage in the following bit of reasoning:

> Everything John says is true. John said *A*. Therefore *A*.

This inference requires a movement from the left-hand side of "'*A*' is true if and only if *A*" to its right-hand side. Similarly, we may want to assert the veracity of someone on the basis of what he has said. So imagine John said only one thing: *A*. Then from *A* we can infer that everything John said is true. Doing so requires a movement from the left-hand side of "'*A*' is true if and only if *A*" to the right-hand side.

Next, I claim that the truth predicate must be *univocal*—that is, that one and only one truth predicate is applicable to all the sentences of our language.[22] Notice that this requirement arises for a really simple reason: any group of sentences, however heterogeneous, provided only that they're sentences of our language, may be asserted by someone (and, consequently, blindly ascribed to by someone else). That is, the univocality of truth doesn't depend on the sentences of our language being any more closely connected than that. In scientific practice, for example, *all* that's required to force the univocality of truth is that we may find, for whatever reason, that a certain set of sentences must be blindly ascribed to *together* (either as true *or* as false). Because no in-principle distinction can be drawn among sentences in our body of beliefs that tells us that now and forever this group of sentences will never be blindly ascribed to under the same description as this other group, we must have a truth predicate that allows us to ascribe commitment to (or denial of commitment to) any sentence of the language without distinction.

Notice how little comes into forcing "true" to apply to *all* the sentences in our body of beliefs: In particular, epistemic considerations about *how* the truth of such sentences are established—how they're confirmed, for example—are irrelevant.

I understand "confirmation holism" is a very weak sense: As I use it, it's the claim that any part of our body of beliefs *may* be brought to bear on any other part, as evidence, as background knowledge—whatever. Other philosophers understand "confirmation holism" perhaps differently: that, in some sense, all the sentences of our body of beliefs are *confirmed* together. If this way of putting the matter amounts to my first formulation, then I'm committed to it. If, however, it means something more, something like there being a well-defined notion of *confirmation* that some sentences (evidential ones) confer on others—so that the confirmational holist is committed to the claim that all sentences in our body of beliefs are confirmed in this special sense *together*—then I'm not committed to it.

---

[22] In my 2000b, I described this requirement as the *topic neutrality* of truth—but *univocality* is perhaps better jargon.

Notice that "confirmation holism" even in my very weak sense *suffices* to compel the univocality of truth, because in bringing one part of our body of beliefs to bear upon another part, blind truth ascriptions may be required. However, as I've just indicated in the preceding few paragraphs, the univocality of truth is forced on us by pragmatic considerations *even if* confirmation holism, *even in* my weak sense, is false.

Concluding Remarks

What's been shown about indispensably applied mathematical doctrine and empirical scientific law? The example of PA is suggestive, and what it suggests is that we're simply committed to the truth of *every* sentence in a body of statements that—as a whole—is characterized by a description (e.g., reference to a schema), which requires blind truth ascription for application, and which is indispensable to empirical science.

This isn't quite true, however. Imagine that (for whatever reason) we refuse to commit ourselves to any implication of PA larger in length than some number n, where n is large enough that no sentence with that number of symbols (or more) can ever be applied by us in an inference. Then, to avail ourselves of an implication *Im* of PA, we needn't engage in the inference: PA is true. PA implies '*Im*'. '*Im*' is true. *Im*. We can say instead: Every implication of PA that is smaller than n is true. '*Im*' is an implication of PA that is smaller than n. '*Im*' is true. *Im*.

What this shows is that if we can descriptively subdivide a set of sentences in an appropriate way, we needn't be committed to the truth of *all* of them, despite having to use blind truth ascriptions. Thus, we needn't be committed to all the sentences in a (schematically presented) theory if we can indicate a subclass of them the truth of which we prefer to be exclusively committed to. Whether this can be turned to the advantage of that philosopher who wishes *not* to be committed to the truth of empirically applied mathematical doctrine or to (some) scientific laws is something I address in chapter 2 (see esp. the discussion in section (B3)).

One last point that should be stressed: Those who are convinced that the ascription of truth to a sentence only amounts to the assertion of the sentence itself may think that escaping the need to use the truth predicate in the case of a class of sentences won't suffice for an escape from a commitment to those sentences (and, *really*, to their *truth*). This is correct, as the discussion of the manifestation condition (also in chapter 2, (B3) will make clear.

2

Circumventing Commitment to Truth despite Empirical Indispensability

Here's where we are: The indispensability of the truth idiom coupled with the indispensability of (certain) statements of mathematics and empirical science seems to require us to assent to the *truth* of such statements in order to apply them empirically. But a number of philosophers are either fictionalistic or agnostic about the truth of empirically applied mathematics, or even about (certain) empirical laws themselves.[1] One can easily imagine strategies to enable such philosophers to retain their respective positions by circumventing, one way or another, the points about blind truth ascription raised in chapter 1. This chapter is dedicated to showing why most of these strategies, which many readers will have in mind, won't work, and why the one that does is insufficient for the purposes of philosophical agnostics and fictionalists.

(A) Attempting to Show Mathematics Is (Actually) Dispensable

Those who want to deny the truth of applied mathematics can attempt to show that such applied mathematics only *appears* to be indispensable; they can try to show that mathematical statements really amount to dis-

---

[1] Examples of those who deny that indispensable scientific laws are true, or at least are agnostic about them: Cartwright 1983, Hacking 1983, van Fraassen 1980. Examples of those who reject the Quine-Putnam indispensability thesis by denying the *truth*, and thus our need to assent to the *truth*, of applicable mathematics: Maddy 1992, Sober 1993.

pensable instrumental inference tickets between empirical statements. Execution of this programmatic claim requires (i) that empirical subject matters cast in a form that commits us to undesirable mathematical abstracta be recast as mathematics-free, put in a form that doesn't commit us to such abstracta; (ii) that the mathematics so applied can be shown to be conservative with respect to the mathematics-free empirical subject matter of (i) (i.e., that the empirical theory plus the mathematical doctrine applied to it doesn't imply empirical statements not already implications of the empirical theory alone); and (iii) that it's *practically possible* to dispense with the inferential role that mathematical statements have on this view.[2]

To date, nothing has been achieved that can be regarded as successfully meeting the challenge of items (i) and (ii), not even with respect to the branch of Newtonian physics that Field 1980 concerns itself with. Item (iii), to my knowledge, has not even been *addressed*. But (iii) looks crucial for reasons already indicated in chapter 1: Suppose headway were made with respect to (i) and (ii). If the resulting mathematics-free empirical theory required a mathematics-free inference practice implementable only by near-immortals, this wouldn't yield the dispensability of applied mathematics. If, despite success in meeting the constraints of items (i) and (ii), scientists still need to routinely assent to (collections of) mathematical statements to ply their trade, this would suffice for their commitment to the truth of such sentences, regardless of the fact that "in principle" such sentences were dispensable.[3]

(B) Attempting to Show Indispensability Doesn't Imply Truth

Some may deny the truth of applied mathematical doctrine or the truth of certain scientific laws *despite* their indispensability. I distinguish three substrategies: (B1) epistemic, (B2) semantic, and (B3) instrumentalist.[4]

(B1) Epistemic

One can argue that the mere indispensability of a statement doesn't suffice for a commitment to its *truth*—one needs something stronger, for example, a certain sort of confirmation or the statement's passing a certain sort of empirical test. So Cartwright (1983), for example, argues that

---

[2] See Field 1980 for an attempt at (i) and (ii) for a restricted bit of physics. See Burgess and Rosen 1997 for a more recent discussion of the technical issues.

[3] There are tricks that finesse the need for (iii), perhaps the need for (ii), and that can weaken (i). See section (B3). Nevertheless, as I show, the execution of even weakened forms of this program still pose insurmountable obstacles.

[4] It's not easy to attribute strategies outlined in sections (B1) and (B3) cleanly to distinct specific thinkers, so in what follows I attribute more than one strategy to the same philosophers.

certain physical laws aren't established as true because consequences aren't derived directly from *them* and tested against experience. Sober (1983) argues that mathematical statements aren't true because they're never tested at all: Statements presumed by the frameworks of empirical tests aren't tested by such tests.

Given the considerations raised in chapter 1, this looks hopeless—it doesn't matter *how* statements indispensable to scientific practice are discovered or established; in particular, they don't have to arise from *any* epistemic process at all—respectable or not. If we discover that we need blind truth ascriptions to certain statements, or that we need to assert the statements themselves, in order to carry out our inferential practices in science, then we're committed to the truth of those statements. Period.

(B2) Semantic

Real Truth, one can argue, isn't an idiom restricted to blind truth ascription as its role; that *deflationary* view of truth is philosophically inadequate. Everyone (let's say) can agree that the Tarski biconditionals are *one* necessary condition on a truth idiom, but *other* conditions are necessary too. Some philosophers may think that one particular additional condition is required; others may claim that there are several competing richer notions of truth. All may agree, however, that unless one or another additional condition is satisfied by indispensable mathematical doctrine or by empirical laws, we shouldn't consider them (really) *TRUE*, although they may be (deflatedly) true.[5]

A variant of this move is to try to claim that such indispensable uses of the truth idiom when applying mathematics or empirical laws are nevertheless nonliteral uses. Thus, although one must *say* that such statements are true, one nevertheless doesn't literally *mean* it.[6]

These won't work either. Let's first consider the richer notion of truth idea and then turn to the nonliteralist ploy. The problem is that any supplementation of deflationist (Tarski-biconditional governed) truth faces a dilemma. Either it falsifies a Tarski biconditional (and so proves unfit for blind ascription), or it fails to be a *genuine* supplementation of the deflationist notion of truth. Let me illustrate these claims with some examples.

---

[5] See Wright 1992 for the view that the Tarski biconditionals provide an unstable minimal notion of truth that *must* be supplemented with additional conditions in order to yield (one or another) richer notion of truth we can use. Candidate additional conditions may include (among others) (i) some sort of correspondence to facts or the world, (ii) verification under ideal circumstances of inquiry, (iii) ultimate acceptance by the community of knowers, and (iv) best coherence with our background beliefs.

[6] My thanks to Sarah McGrath for urging me to consider this objection. Attributions of metaphoricality or pretense to avoid otherwise apparent commitment to the truth of certain sorts of statements or to the existence of certain sorts of entities, e.g., fictional or mathematical ones, are becoming popular. I further discuss these views, and my objections to them, in chapter 3.

First, consider the possibility of supplementing the notion of truth with the constraint of warranted assertability:

> '*A*' is TRUE iff *A and*
> '*A*' is TRUE if '*A*' is warrantedly assertible.

From these conditions, it follows that:

> *A* if '*A*' is warrantedly assertible.

In particular, the rug is green if "The rug is green" is warrantedly assertible. But this is ruled out by the following kind of sentence (which we intuitively accept):

> "The rug is green" might be warrantedly assertible, even though the rug isn't green.

That is, our intuitive acceptance of this "might" statement shows that TRUTH so understood can falsify Tarski biconditionals. Similar sorts of modal statements can be used to disable any supplementation of the notion of truth that is, as previously noted, epistemic in nature.[7]

However, there seem to be ways of supplementing the Tarski biconditionals that don't generate such problematical modal statements. Consider the suggestion that what's called for is a substantial notion of truth that replaces the right-hand side of the Tarski biconditionals with what amounts to a compositional analysis of the meaning of the sentence, for example:

> (COM) "Snow is white" is TRUE iff "Snow" picks out an object *o* and "white" picks out a property *W*, and *o* has the property *W*.[8]

It doesn't seem right to say in this case either that

> It might be that snow is white even though it's not the case that "Snow" picks out an object *o* and "white" picks out a property *W*, and *o* has the property *W*,

or that

> It might be that it's not the case that snow is white even though "Snow" picks out an object *o* and "white" picks out a property *W*, and *o* has the property *W*.

Nevertheless, there are troubles if the idea is that what (COM) supplies is a notion of TRUE richer than the deflationist notion of true. There are two possibilities here. The first is that the right-hand side of

---

[7] In particular, (ii), (iii), and (iv) of n. 5 are examples. See my 2000b, pt. 2, § 7, for further details on this argument and for consequences.

[8] My thanks to Penelope Maddy for urging me to explicitly discuss the compositional approach to truth. I should add that my argument against supplementations of the sheer quantificatory role of truth appears (in its essentials) in my 2000b, p. 51 (see (ii), in particular).

(COM) goes strictly beyond the right-hand side of the corresponding Tarski biconditional in the case of *certain* sentences. Imagine, for example, that a nominalist argues that although "1 is a number" is true, it's not TRUE because there is no object o picked out by "1," and no property W picked out by "a number" so that o has the property W. In this case, it immediately follows that TRUE doesn't satisfy (all) the Tarski biconditionals, since "'1 is a number' is TRUE iff 1 is a number" is false.

The second possibility, obviously, is to have TRUE satisfy (all) the Tarski biconditionals. But in that case, for want of contrast, the right-hand sides of the instances of (COM) are toothless: as far as *truth* is concerned, they add nothing more to the notion above and beyond the Tarski biconditionals. Why? Simply because the supplemented notion of TRUE applies to *every* sentence of the language regardless of whether we take that sentence to have genuinely referring terms in it or not; in particular, given that we're required to assert that "1 is a number," we're similarly required to assert that "'1' picks out an object o and 'number' picks out a property W, and o has the property W."[9]

If the foregoing is right, a similar objection applies to the nonliteralist ploy. Nonliteralists want to claim that, despite the fact that their use of the truth idiom involves Tarski biconditionals, nevertheless nonliteralists shouldn't be taken as literally making a truth ascription. Given that *any* truth idiom must obey the Tarski biconditionals, the only option for nonliteralists is to claim that there is *something more* to the concept of truth, and that this something more is needed for a literal assertion of truth. But now the same objection can be made against the nonliteralist that was already raised against that philosopher who thinks the notion of truth involves more than what's required by its deflationary role—any such notion of truth that genuinely goes beyond the Tarski biconditionals can't be used for blind ascription.

(B3) Instrumentalist

This last strategy is one I spend a lot of time on (pretty much the rest of this chapter) because it's a strategy that attracted many early practitioners

---

[9] Notice that this is *not* an objection to compositional analyses of language per se, or to the intimate connection such analyses, following Tarski, have had to the notion of truth. Any axiomatization of any notion can be *supplemented*—we can always add new terminology to the language and then connect that terminology to the old language in a way that doesn't disturb the previous axioms; that's not the issue. The point is that the Tarski biconditionals can't be supplemented in a way that can legitimately be described as enriching the notion of *truth* (vs. merely wedding other concepts to it). Compositional analyses of language don't themselves illuminate the notion of truth, and, related to this, attempts to wed notions of truth to ontology via compositional analyses of language are misguided. The second point will be illustrated in what follows. For more details on the first point, see my 2001 and my forthcoming(b).

One caveat: I've again suppressed refinements about the truth predicate in the interests of expository ease. See the general remarks about truth that chapter 1 closes with, esp. n. 21.

of philosophy of science. This is in part because scientists themselves sometimes take an instrumentalist attitude toward theories they nevertheless apply.

It may be argued that although blind ascriptions of some sort are required for scientific practice, it's not that what's needed are blind ascriptions of *truth*—at least not blind truth ascriptions to the *whole* theory. All that's needed are blind truth ascriptions to whatever sentences are actually needed for empirical application; the other (inferential) machinery of the theory needn't have truth ascribed to it at all.[10] Indeed, it can be claimed that this instrumentalist take on scientific theories is routine in science itself.

Ptolemaic astronomy, for example, is regularly offered as a paradigmatic historical illustration of how a scientific theory can be understood purely instrumentally while still being empirically *indispensable*.[11] Suppose, therefore, that a certain theory P implies consequences about the apparent movements of the planets and the luminaries over time. Call these consequences the *apparent-motion consequences of P*. We can take the apparent-motion consequences of P to be true without taking anything P implies about the *real* motions (of the planets and luminaries) to be true. Even if P isn't finitely axiomatizable, the following inference is acceptable: The apparent-motion consequences of P are true; 'Im' is an apparent-motion consequence of P; 'Im' is true. Therefore Im.

At the moment, I'm not concerned with evaluating the instrumentalist strategy as used by philosophers to avoid realist assumptions—I get to that shortly. Instead, I'm granting that scientists often take an instrumentalist attitude toward their theories, and I'm addressing the question of exactly what's involved when they do so—how, that is, we can identify when a theory is being treated instrumentally. Once I complete this analysis, I turn to the question of whether instrumentally inclined philosophers can use this strategy more broadly.

Some philosophers may be tempted to think that if a theory is empirically indispensable, then one has *no choice* but to accept its truth. That

---

[10] The reader will notice in what follows that I'm not distinguishing between taking an empirically applied (and, sometimes, indispensable) scientific theory to be truth-valueless and taking it to be false. For our purposes, this particular distinction doesn't matter. So hereon, both applied empirical theories that are taken to be false and those that are taken to be truth-valueless are described as "instrumentally construed."

[11] What makes Ptolemaic astronomy empirically indispensable (at a particular time) is that at that time it offers the only way to predict and retrodict the apparent movements of the planets, the sun, and the moon. Many philosophers, following Duhem, think Ptolemaic astronomy was understood instrumentally in just this way by its practitioners. Hatfield (1990) writes sternly: "Much mischief has been caused by the acceptance of Duhem's characterization of Ptolemaic astronomy as inherently 'instrumental' . . . " (p. 155, n. 37). To avoid historical impropriety, I subsequently substitute the ahistorical example of P, a theory used instrumentally by its (imagined) practitioners, for actual Ptolomaic astronomy as historically practiced. It certainly was *possible* for Ptolomaic astronomy to be treated instrumentally, and besides, there are other examples of theories that *are* treated instrumentally in current science, as I indicate later.

is, the only way to escape a commitment to the truth of an empirically indispensable theory is to engage in *paraphrase*: replace that theory with *another* that has the same empirical role as the theory it has replaced.

This is just *wrong*: A theory, even an empirically indispensable one, can be treated instrumentally (under certain conditions), as the *P* example just showed. What's needed is an implicit response to a question we must answer in order for the attribution of an instrumentalist attitude toward an empirically indispensable theory to be cogent: How does the difference in attitude toward sentences we take to be true and toward sentences we take instrumentally *manifest* itself? The answer is that if a theory (or sentence) is one we think true, then we must also include in our body of beliefs (at least implicitly) *all* of its implications. In *P*'s case, if we really do take the theory *only* as an instrumental device for recognizing the apparent motions of celestial bodies, then we ignore or *quarantine* the other implications of the theory from our communal body of beliefs.[12] That is, we don't draw implications from the theory with respect to the real motions of celestial bodies—in particular, we don't do so when concerned with evaluating what evidence we have or could have for P.[13]

If our commitment to the truth of a sentence comes only to a commitment to that very sentence itself, there's a puzzle about how a sentence can be *useful to* our body of beliefs and yet *not* be regarded as *true*. For to include it *in* our body of beliefs—given the Tarski biconditionals—*is* to regard it as true. The solution I've offered is that if a sentence *really* is included in our body of beliefs, all its implications must be included, as well. Thus, a necessary condition on a sentence not being included in our body of beliefs (except as an instrumental tool for codifying a *sub*class of its implications that *are* included) is that *not all* its implications are so included.

The scientific insight often at work in the application of false theories is that a false theory can have *true* implications (even though it won't be that *all* its implications are true) and these true implications can not only be useful but also, in certain circumstances, be empirically indispensable (because application of the true theory of a certain phenomenon is intractable or unknown).[14] A simple example of an instrumentally construed

---

[12] I use "quarantine" with some misgivings. I like its sound, and its connotations, except for the suggestion (for some readers) that all quarantines are temporary. They needn't be in this case. Although some theories may shift in their status from an instrumental tool to something we whole-heartedly believe, in many cases this isn't possible, and the implications of the theory placed in quarantine remain there forever. ("Exile," an earlier option I considered, is missing the connotation of temporality but sometimes gives rise to awkward phrasing; "ostracize" is just too ugly. So "quarantine" it is.)

[13] I've previously (1997a, pp. 196–7; 2000b, p. 54) discussed this under the rubric of "isolating a theory understood instrumentally from the rest of our web of beliefs."

[14] A special case of a theory with quarantined implications (which may be the most common case in practice) is where *none* of the implications of that theory are contained in our body of beliefs; such cases, where none of the implications of a theory are taken as true but nevertheless they're still seen as useful, is when such implications numerically *approximate* what's true (relative, of course, to a background theory that we *do* take to be true).

theory with these properties can be found in an introductory physics text (Purcell 1985):

> Consider ... a slab of material of thickness dz, sliced out perpendicular to the direction of magnetization. ... The slab can be divided into little tiles. One such tile, which has a top surface of area da, contains a total dipole moment amounting to $M\,da\,dz$, since M is the dipole moment per unit volume. ... The magnetic field this tile produces at all *distant* points—distant compared to the size of the tile—is just that of any dipole with the same magnetic moment. We could construct a dipole of that strength by bending a conducting ribbon of width dz into the shape of the tile, and sending around this loop a current $I = Mc\,dz$. ... That will give the loop a dipole moment:
>
> $$m = I/c \times \text{area} = (Mc\,dz)/c \times da = M\,da\,dz$$
>
> which is the same as that of the tile. (p. 423)

Notice that at *distant* points this particular computational device yields the same magnetic field as the tile's real microstructure yields. But that this "ribbon theory" is purely instrumental is shown by the fact that it can only be used to compute the tile's magnetic field at distant points. The "presence" of such ribbons can't be used to discover other properties of the tile, for example, the magnetic field at near points, the mass of the magnetic material, and so on.[15]

I've stated the quarantining of (some of the) implications of a theory as a necessary condition for treating that theory instrumentally, but I'm willing to go out on a limb here and describe it as both a necessary *and* a sufficient condition: How else are what we take to be differences in truth-bearing status for sentences to manifest themselves if not via differences in their inferential roles? For an attitude of instrumentality toward a theory (vs. a commitment to its truth) *must* manifest itself in some way other than by the *mere* expression of an *attitude*: If a theory and its implications are treated in exactly the same way as other theories we believe in, if its sentences are used in deductions, in explanations, in reductios—when possible—just as the sentences of its fellows are, then an attitude of instrumentality toward it (vs. its fellows) amounts merely to the *claim* that we take a different attitude toward the theory (that we turn up our noses at it when using it), *without* that attitude emerging in explicit scientific practice. Such free-floating attitudes can't be taken seriously.

However, it's important to distinguish the kind of case where we disbelieve in a theory and as a result ignore (certain) implications of it

---

[15] Presumably, this approach is indispensable if we want to make predictions about the magnetic field at distant points from an arbitrary piece of material of this sort. This is because of, in most cases, our inability to compute the actual field generated by the slab's microstructure. Nevertheless, it's perfectly clear that it's entirely legitimate for scientists to take an instrumental attitude toward this theory.

from the very different cases where we ignore (certain) implications of a theory either because they're useless or trivial or because of epistemic reasons: We can't apply or test those implications (because they describe possibilities that are—currently—out of reach for us). For example, had P been taken as true, the apparent motions of the planets would have been understood to be due to their real motions as presented by P, and so, should our access to evidence for the real motions have then changed, we would have then tried to determine whether there was or wasn't evidence for such motions. We wouldn't bother doing this if P were construed instrumentally.[16]

We also want to distinguish the quarantining of implications from the sort of case where a theory *seems* to have an implication that's refuted experimentally (or "by experience"). Of course, many things can go wrong during the long route from theory to experiment, and we may not know what to blame or how to explain what's going on. Here, again, when we seem to ignore an implication of a theory that seems to have evidence against it, it's a matter of epistemology—not a matter of that implication being in quarantine.

Lakatos once described all scientific theories as "born refuted": There are always apparent exceptions to them. Because there is no deductive straight line from a theory to experimental or observational implications, this melodramatic sentiment doesn't quite present the correct picture. Nevertheless, although apparent anomalies can be ignored for years, still this isn't a matter of the placing of implications in quarantine: the latter only happens when we have reasons to think the implications are false or when we really haven't any reason to think the theory—outside its computational value for certain purposes—is either true or false. An example is when we blatantly don't *bother* to deduce certain implications of a theory (they're not relevant to its computational value), not when we've "deduced" an anomaly that no one knows how to explain.

The reader may think that putting some implications of a theory in quarantine can't be a necessary condition for taking that theory instrumentally. For example—so the objection goes—a sentence can be *in* our body of beliefs (in as robust a sense as any other sentence is) and yet not be something we take *literally* true. I'll have more to say about this suggestion in chapter 3, specifically with respect to its application to the semantics of discourse about fiction. Let me state now, however, that one symptom of the metaphorical use of a word in a sentence is that at least some (and, often, all) of the (strict) implications of that sentence *are* placed in quarantine. This is clearest with idioms such as "He kicked the bucket," but it's even true of genuine metaphors in song lyrics and po-

---

[16] This is important. The potential evidence for or against a theory is always subject to change because certain implications of a theory (heretofore insulated against confirmation or infirmation) can become testable via mathematical breakthroughs or instrumental advances. I put this point to polemical use a few pages hence.

etry. Consider the old Bob Dylan line: "The sun's not yellow; it's chicken." (Surely we don't take all, and perhaps *any*, of the literal implications of this sentence to be true, and this is how we intuitively recognize the presence of metaphoricality in this case.) Thus, it's reasonable to assert that a metaphorically construed sentence is *not* part of our body of beliefs; its literal paraphrase (if it has one) *is* part of our body of beliefs; and even if it hasn't a literal paraphrase, still, some of its implications *may* (literally) be part of our body of beliefs.[17]

Three caveats: First, of course, it's always possible for the status of a theory to change—an instrumental theory may be one we later take to be true (or, more commonly, vice versa). And this change in attitude will be due to our (now) concerning ourselves with implications of the theory that we once ignored.

Second, there is one way of taking a theory instrumentally that I should be explicit about. This is when a *paraphrase* for it is available (and so it's only out of sheer laziness, as it were, that we use the unparaphrased theory). I'm not explicitly concerning myself with this sort of case (and so I'm not qualifying my claim that placing (some) implications of a theory in quarantine provides both necessary and sufficient conditions for a theory to be instrumentally construed because I'm implicitly treating a theory as empirically indispensable when *no* such paraphrase is available). It's worth pointing out, however, that it's well known to philosophers that there are scientific theories that scientists seem to treat instrumentally but for which no paraphrase seems available. Maddy (1997) tells us that on the paraphrase approach:

> [A] claim about a continuous fluid would be replaced by, or translated as, a claim about what happens to actual fluids as they approximate ever more closely to ideal fluids, that is . . . as their molecules become ever more tightly packed together, approximating continuous matter. But this is all wrong. If the molecules of an actual fluid are packed tightly enough, it stops being a fluid, and even if it didn't, the best we could approximate in this way would be density, not full continuity. (p. 145)[18]

---

[17] There's more to say about this interesting topic but I can't get into it in *this* book. One point that needs stressing is that placing some implications of a sentence or theory in quarantine is hardly a *sufficient* condition for the presence of metaphor: The quarantined implications of instrumentally construed scientific theories hardly show *those* theories to be metaphorical.

[18] These particular considerations are hardly *fatal* to the paraphrase proponent—for whoever said that the only way to paraphrase (and thus eliminate) a scientific theory was to show that it approximates theories that we take as really true, in the sense that a limit operation on the entities (presupposed in the approximating theory) that takes *them* to real entities similarly takes the approximating theory to the *true* theory. (Well, there may be places in Quine's writings where *he*, at least, seemed to *imply* something like this. But he shouldn't have.) But in any case, Maddy is right about the prospects of the paraphrase option: scientists are perfectly willing to use any theory at all—whether they believe it or not—if it can be shown to have (some) empirically valuable implications. They don't stop to show that the theory so used (and yet instrumentally construed) is paraphrasable away.

Notice by contrast how easily this case is dealt with on the quarantine view of instrumentally construed empirical theories.

Third, suppose we had *no* truth idiom and no way of expressing blind ascriptions. Even so, we could *only* mark a difference between sentences we were committed to and ones we were merely taking instrumentally by placing (some) implications of the instrumentally construed sentences in quarantine. For someone who sees facilitating blind truth ascription as the sole purpose of the truth idiom, *this* is the heart of the indispensability argument. As I've shown, if it's claimed that some sentence in our body of beliefs—although (1) ineliminable by paraphrase and (2) none of its implications are in quarantine—is *still* not to be taken as *literally true* (see, e.g., Yablo 2000 for this kind of view), the response is that if *no* implication of the sentence—literally construed—is in quarantine, then the claim that this sentence is not literally taken to be true has *no* manifestation whatsoever with respect to the role of that sentence in our body of beliefs. (Offstage whisperings about instrumentality or metaphoricality don't count as "manifestations.")[19]

Let's turn now to philosophical attempts to attribute instrumental status to scientific theories. The threat of an alleged instrumental attitude toward a theory evaporating into the ineffable seems to sour a (traditional) instrumental approach to theories—to introduce a division in the vocabulary of the language (theoretical vs. observational) that is then to impose a separation of the empirically valuable observational truth-bearing consequences of a theory from the rest of its purely theoretical consequences; the latter can then be treated as truth-valueless. Many philosophers have pointed out that this project is doomed if only because the required separation of vocabulary into observational and theoretical terminology can't be carried out;[20] this is right, of course, but I'm pressing a *different* concern, which is that such a distinction in vocabulary is taken to *support* a distinction in attitude toward sentences we take to be true and those we treat merely as instruments for deducing something else we take to be true. But such a difference in attitude can't reside *solely* in nomenclature: It must be grounded in a difference in how the terminologically distinct sentences are treated or used in our body of beliefs to avoid the charge that the purported difference in attitude involves (at most) an ironic stance toward vocabulary—in particular, as I pointed out

---

[19] The considerations just raised reverse a claim made in my 2000b, pp. 50–1. Also see the second-to-last paragraph of this chapter.

The reader may think there are still other ways to manifest an instrumental attitude toward theories. I'm skeptical that any of these possibilities will prove valuable in the way that philosophers who urge an instrumentalist attitude toward scientific law or mathematical doctrine need; see the discussion later of Cartwright (1983), where I raise another such possibility.

[20] For classical animadversions to the observation/theory distinction as applied to terms, see, e.g., Putnam 1962, Maxwell 1962, or Achinstein 1965.

earlier, (certain) implications of (some) of the sentences containing theoretical vocabulary must be placed in quarantine.

It might be thought that what's required is (in principle, anyway) available to those who believe in the observation/theory distinction for terms—observation sentences (on this view) are confirmed and infirmed in a strikingly different way from theoretical sentences: sheer observation does the job for observation sentences; theoretical sentences, on the other hand, only receive confirmation or infirmation inferentially via the observation sentences observed to be true or not; that is, not *all* of their implications are of interest, only the observational ones.

A more recent version of the same suggestion is offered by Cartwright (1983), who argues that many scientific theories aren't taken (or should not be taken) to be true because they themselves aren't directly applied empirically or directly tested. Instead, she claims, only lower-level theories (phenomenological theories) are so applied and tested, and only these we *do* (and should) take to be true.

My objection to this version of the traditional observation/theory distinction, and to Cartwright's view, starts by recollecting the distinction between the case where a theory is taken instrumentally (and so there are implications of the theory in quarantine) and the very common but different case where the theory is taken to be *true* but, because of reasons of intractability, certain implications of the theory haven't been drawn or tested. The latter case involves circumstantial epistemic difficulties that aren't indicators of our taking an instrumental attitude toward a theory.

So, for example, were our observational powers enhanced, what was previously described as theoretical terminology would be reclassified as observational. Were our powers of discrimination, for example, sufficiently sharpened, we would be able to see the movements of (large) molecules, and thus the truth or falsity of previously theoretical sentences about the trajectories of such molecules would become matters of observation. In the same way, it can be objected to Cartwright that many theories can't be directly applied to their proper domain of application because of sheer mathematical intractability or because of the complexity of the situation the theory is to be applied to. When, however, an implication *becomes* deducible or applicable because its applicational intractability has been surmounted in some way, the real attitude of scientists toward the theory is revealed: The implication is immediately exploited. Were the theory really instrumentally construed, the implication would be ignored instead. The point is that as long as we're aware that our evidential basis for recognizing the truth or falsity of sentences can change, even if that basis is taken to be purely observational, we can't regard ourselves as treating (certain) sentences, say, as neither true nor false merely because we can't use them (yet) in evidential arguments or in applications.

It may be thought that these are unfair objections both to those traditional philosophers committed to grounding an instrumental attitude toward theories on the basis of an observation/theory distinction, and to those like Cartwright (1983) who think inability to deduce testable consequences from a theory consigns it to instrumental status: One might claim that there are other ways we can manifest an instrumental attitude toward theories apart from putting some of their implications in quarantine. One can tentatively adopt a theory and draw out its implications but not commit oneself to it until genuine evidence in its favor comes along.[21] Perhaps this is the sort of model for instrumentalist attitudes that Cartwright (1983) has in mind: Insufficiently confirmed theories are ones that we have—at best—only tentatively adopted. Similarly, it can be argued that one can only (at best) tentatively adopt mathematical statements—because these are never confirmed or tested at all.

Unfortunately for this suggested defense, these claims just aren't true to the phenomenology of tentative belief adoption. If one whole-heartedly uses *all* the implications of a theory (that one can), intermixing the implications of one's "tentatively" adopted theory with the other sentences in one's body of beliefs, no sense remains to the claim that the theory has been (only) tentatively adopted. *Real* hesitation about commitment to a theory leads to a *pretense* of commitment: one draws implications from it to see what those implications are (to see what one is getting into); certainly, one does *not* rely on it when (in particular) doing actual science (that would be too risky).

Qualifications: One may believe that a theory is only approximate in some way and apply *it* empirically to get data that tell us what the true theory is (or at least moves us a few stages *nearer* to the true theory). And using the approximate theory this way may be the only way *to* the true theory. (For illustrations of this method of "successive approximations" in Newton's work, see Smith 2000 and 2002.)

One must be careful when considering these sorts of possibilities. If the theory taken to be true isn't in hand, but the "approximating theory" is contained within our body of beliefs to the same extent as any other theory, we're (temporarily, perhaps) *committed* to the truth of the so-called approximate theory although we don't *expect* to be committed to it for very long. (This is a prediction, of course, and we could be wrong.) If, however, (1) the approximating theory TH is *only* used evidentially to provide data that will be used to overthrow it, then because other implications of TH are in quarantine, it *is* being treated instrumentally. Also, (2) if TH is *generally* applied empirically, but there is another theory in place in our body of beliefs against which the approximate truth of TH can (in principle) be measured numerically—so that, for example, a concern with how inaccuracies due to the approximate nature of TH are invariably involved in the application of TH—then TH is being treated instrumentally. In this second case, it can be—as I've indicated before—that we can't *write down* the theory that we take to be true, although in relation

---

[21] I'm going along (for the sake of argument) with this psychologistic take on commonly held beliefs, but see the second paragraph of n. 13 of the introduction.

to it, we can describe what we *do* write down as approximate (and, further, specify how it's approximate).

What's at work helping to distinguish these cases is the simple principle that we *can't* step out of our own body of beliefs. We *can* believe something to be true that we can't write down; something, that is, can belong to our body of beliefs without our being able to express it explicitly (say what it is). And in relation to such inexpressible theories, what we can write down may be regarded as *approximate*.

But it's wrong to describe a theory as approximate on the basis of a *hunch* that applying it will yield data that will falsify it (and enable us to (1) replace it with a theory we can regard as true, and (2) regard the first theory as approximating). *Once* this happens, that's fine. But *hopes* for new theories aren't new theories.

This even applies to a case where we find ourselves committed to a pair of theories both of which are indispensable, and all the implications of which are contained in our body of beliefs, but which we (1) want to unify and (2) recognize *can't* be unified in their current forms. Even here, which seems to be the sort of case we presently face with general relativity and quantum mechanics, it's not correct to describe ourselves as committed to their successor theory, and *not* them. Again, *hopes* for a successor theory aren't the *presence* of a successor theory; unless we can quarantine some or all of the implications of the theories we were previously committed to the truth of (either by treating some of their implications as only approximate with respect to the new theory or by explicitly quarantining some of their implications even without a successor theory), we must take those theories to be true *tout court*.

Of course, the same is true of a single theory, quantum mechanics, say, which we may agree isn't suitable as it stands. Nevertheless, because all its implications (that we can currently derive and test) have proven to be experimentally accurate to within extremely small thresholds—as small as we're currently capable of testing it to—we're committed to its truth until we replace it or find implications of it that we can legitimately quarantine.

One last point about the instrumentalist strategy: It allows (at least in principle) two less arduous versions of the strategy outlined in section (A), as I indicated in footnote 3.[22]

(A1): Presume that we intend all applications of a mathematical theory M to an empirical subject matter to function as inference tickets from

---

[22] Recall that strategy described in section (A) is the attempt to show that applied mathematics only appears to be indispensable, and that mathematical statements are actually dispensable instrumental inference tickets between empirical statements by (i) recasting empirical subject matters, which commit us to undesirable mathematical abstracta, so that they don't so commit us; (ii) showing that the mathematics so applied is conservative with respect to the mathematics-free empirical subject matter of (i); and (iii) showing that it's practically possible to dispense with the inferential role that mathematics has on this view.

(mathematics-free) empirical statements to other (mathematics-free) empirical statements. Then, if we can execute (i) (recast the empirical subject matter in a mathematics-free way S) and execute (ii) (show that M is conservative with respect to S), we can bypass the requirement that the truth predicate apply to M like so: The mathematics-free implications of S are true. 'Im' is a mathematics-free implication of $M + S$. Therefore (by (ii)) 'Im' is a mathematics-free implication of S. Therefore 'Im' is true. Therefore: Im. Thus, requirement (iii) lapses.

Suppose (ii) isn't to be had. ($M + S$, although syntactically consistent, is *not* conservative over S.) Is that a problem? It's not clear why it should be. S has to be incomplete in such a case (there must be sentences s_1 in the language of S with neither s_1 nor $\neg s_1$ deducible from S). Can M still be described as an inference ticket from sentences of S to sentences of S? Yes, if we choose the *intended models* of S so that all the implications of $M + S$ hold in every intended model of S. And this can always be done.[23] So (ii) lapses as a requirement, as well. Needless to say, the program requiring just (i) has no more chance of success than the more stringent program discussed in section (A).

(A2): Suppose we want to consider $M + S$, as a whole, to be an instrumental inference ticket *to* its mathematics-free implications. In this case, (i) and (ii) lapse as well as (iii): We don't need to recast the empirical subject matter of $M + S$ in a mathematics-free way S. Instead, we only require a weakened (i'): We can cast those implications of $M + S$ that we need to apply or test into a mathematics-free form. It turns out that even this (very) weakened program has no more of a chance of success than does the original program discussed in section (A).

This is perhaps not obvious. After all, to nominalize *all* of physics looks difficult—but suppose that we want physics only for the purpose of deducing the motions of large macro-objects such as boulders careening through space. Why wouldn't the needed implications about the movements of boulders *already* be nominalized? This is a good question; in the last paragraph of chapter 9, I will show why this approach isn't desirable.

Apart from this, I should note that those like Field (1980), who want an epistemic *justification* for the application of mathematics in terms of its conservativeness with respect to nominalized versions of the empirical subject matters it's applied to, won't find this second suggestion described in section (A2) appealing.

I turn now to an observation about an argument that philosophers prone to dismiss the truth of (certain) empirical laws often raise. Both

---

[23] This is obvious if the logic in which these theories are couched is first order. But what if it's second order? Then it still can be done—but in such a case, because the *logic* is incomplete, we have to treat the alternatives not as alternative models of S but as alternative semantic frameworks for second-order logics (with different principles). This is as far as I can get into this now.

Hacking (1983) and Cartwright (1983) point out that more than one scientific theory may be applicable to the same phenomenon. In studying the electron, say, there are any number of competing theories—which disagree—that scientists may apply to yield results about electrons. This motivates the claim that, *according to scientific practice*, although the electron is taken to be real, theories about electrons aren't generally taken to be true.

I've two objections to the foregoing. First, I disagree with the description given of scientific practice: One must distinguish between theories recognized *by scientists* to be only approximating and ones not so understood. Approximating theories (ones idealizing away from one or another aspect of the phenomena they're applied to) can approximate *in different ways*. Because of this, they can give different answers when applied to the same phenomenon. But this hardly shows a commitment to inconsistent theories.

But, second, the considerations about blind truth ascriptions show that *even if* scientists were committed to contradictory theories, one could not blithely gloss this as a matter of the theories not being taken to be *true*. That would be to mistake why the truth idiom is involved in scientific practice to begin with. The right move—if a philosopher insists on the collective inconsistency of scientific doctrine—is to focus on the *logic* used in science, to argue that scientists are using one or another paraconsistent logic, for example, instead of a classical one.

As I've said, I prefer to leave the logic alone and take a closer look at how the so-called competing theories are justified in applications, in particular, to locate what more general theory they're being culled from, and in which ways they're differing approximations of that theory. Of course, I'm making an empirical claim here: It could turn out that scientists *really are* committed to inconsistent theories in the way Hacking and Cartwright suggest. But I doubt it. And further, I suspect that if this could be shown, scientists would be *really* upset. After all, as I've already mentioned, the *apparent* incompatibility of general relativity with quantum mechanics already makes (many) physicists uncomfortable. It's this perceived incompatibility that is the primary factor driving so much of the research toward a unified theory of "everything."

Let me stress explicitly an implication of my last paragraph: Attribution of an instrumental attitude toward a scientific theory is an *empirical matter* on my view, and one that doesn't turn solely on what a scientist or group of scientists happens to *claim*. Further, this means that attributions of instrumentality toward indispensable scientific theories aren't matters for philosophers to stipulate (on the grounds of one or another philosophical program). Instead, an instrumental attitude toward such a theory must manifest itself the way I illustrated earlier: (some) implications of that theory must be placed in quarantine.

Two last points: First, this take on how attributions of instrumentality are to be recognized allows the empirical possibility of a science where we're committed to the truth of none of our (scientific) theories (*all* of them have implications in quarantine). I see no reason, however, to think current science is like this. Current science is still in the business of searching for laws that apply *without reservations* to the world.[24]

Perhaps this claim is more controversial than I realize. Truesdell (1979) writes: "A theory is not gospel to be believed and sworn upon as an article of faith, nor must two different and seemingly contradictory theories battle each other to the death. A theory is a mathematical model for an aspect of nature. One good theory extracts and exaggerates some facets of the truth. Another good theory may idealize other facets" (pp. 72–3). The context of this quote does *not* suggest that Truesdell is making a claim about specific *types* of scientific theories versus scientific theories *in general*. Notice how the quarantine view successfully explicates Truesdell's way of characterizing the instrumental status of such theories. However, he gives no reason, as far as I can see (and despite his apparent claim), for thinking that *all* physical theories are like this.

Second, Maddy (1994, 1997) has tried to undermine the assertion that the indispensability of applied mathematics forces us to regard it as true by exploiting an apparent analogy with indispensable empirical theories that scientists nevertheless take an instrumental attitude toward.[25] In

---

[24] Something stronger can be said: Scientists are still in the business of providing *explanations*, and they can't do *that* without *true* theories. See my 2000b, p. 61, where I argue for this. Also see Levin 1984 (p. 127), where the point that explanations must be true is asserted, and Hempel 1948 (p. 248), where an adequacy condition of something being a "sound" explanation is that it be true.

Cartwright (1983) offers an interesting objection to the claim that explanations must be true that she attributes to Duhem and van Fraassen. She writes: "Duhem and van Fraassen take truth to be an external characteristic to explanation. Here is an analogy. I ask you to tell me an interesting story, and you do so. I may add that the story should be true. But if I do so, that is a new, additional requirement. . . . " She adds, a few lines later: "Van Fraassen and Duhem challenge us to tell what is special about the explanatory relation. Why does the truth of the second relatum guarantee the truth of the first?" (pp. 90–1).

Those who see truth as I do, i.e., as simply a device by which commitment to sentences is articulated when direct assertion of the sentence is otherwise awkward, deny that the truth of a story is a new additional requirement on the story (except insofar as it's a requirement that the story be told sincerely and declaratively—e.g., not fictionally, or within scare quotes or in pretense). So, too, Cartwright's question about what's so special about the explanatory relation is one to be *deflected*, not answered: The truth of the second relatum isn't supposed to guarantee the truth of the first; rather, if the first isn't true, then the relation between them fails to be the explanatory relation. (Compare: What's so special about implication that the truth of the consequence is guaranteed by the truth of the antecedent? Nothing except that if this isn't the case, then the supposed implication-relation between the two sentences isn't one.)

[25] Caveat: Maddy's discussion is concerned both with the truth of indispensable doctrine and with a consequent commitment to the entities posited by such doctrine, and so considerations about commitment to entities and about commitment to truths are intertwined in her discussion. I think these must be carefully separated, as the analysis in this book should make clear. Maddy (1997), upon concluding that many indispensable empirical theories are nevertheless regarded as false (contrary to Quinean dicta), does suggest that "we might place blame on the criterion of ontological commitment; we might deny that

her 1997 (pp. 143–4), for example, she notes that many empirical theories are taken to be (literally) false.

I've already given a couple of examples of such, but they're numerous, and Maddy alludes to many more:

> [W]e treat a section of the earth's surface as flat, rather than curved, when we compute trajectories; we assume the ocean to be infinitely deep when we analyse the waves on its surface; we use continuous functions to represent quantities like energy, charge, and angular momentum, which we know to be quantized; we take liquids to be continuous substances in fluid dynamics, despite atomic theory. (p. 143)

She continues: "On the face of it, an indispensability argument based on such an application of mathematics in science would be laughable: should we believe in the infinite because it plays an indispensable role in our best scientific account of water waves?" And she concludes with this: "Given that science abounds with idealizations, and given that we are not justified in drawing ontological conclusions from an application of mathematics that occurs in such a context, we should turn our attention to the role of mathematics outside explicit idealizations" (p. 146).

But this conclusion is too swift if the suggestion is that such applications of mathematics can simply be seen as false in the same way that the empirical doctrine it's applied with is seen as false.[26] As Resnik (1995) has already pointed out, the usefulness of false applicable empirical theories requires that the mathematics employed with such theories be true.[27] I would only supplement this point with the observation that, in light of the foregoing discussion, implications of applied mathematical doctrine are never placed in quarantine in the way that (instrumentally construed) implications of empirical doctrine are. On the contrary: If a tabletop fails to be perfectly flat, it's not that we quarantine *certain* Euclidean theorems; we actually *use* those theorems to determine *to what degree* the tabletop fails to be Euclidean. We may, of course, choose not to apply Euclidean geometry to the tabletop *at all*—its shape may obviously dic-

---

we are committed to everything our theory says 'there is'. Perhaps the notion that all existence claims are on the same footing—the 'univocality of "there is"' as Quine calls it—is inaccurate as a reflection of the function of scientific language" (p. 143); but her immediately following suggestion that one should perhaps consider pre-Perrin "there are atoms" as fictional and post-Perrin "there are atoms" as literal indicates that, in any case, she sees a strong analogy between how scientists deny (or reserve judgment on) the existence of empirical entities, and similar instrumental attitudes toward mathematical objects. It's this purported analogy that I primarily want to challenge. See my 1997b, or better yet (because the reader presumably has this book in hand), see chapters 3, 4, and 5. For purposes of the immediate discussion, I'm focusing only on the suggestion that because scientists cogently take indispensable empirical doctrine to be false, there is hope that something similar can be done with indispensable mathematical doctrine.

[26] See n. 25.

[27] Maddy (1997, p. 153, n. 32) takes note of Resnik's objection but describes herself as "unconvinced" without explaining why.

tate the application of some non-Euclidean geometry to analyze the properties of figures drawn on the tabletop. But this is *not* to quarantine the truths of Euclidean geometry. They simply don't apply; they aren't falsified. (I.e., the point about mathematical truths not being quarantined simply reduces to the old point that mathematical truths—contrary to Quinean hopes—are *not* falsified by empirical application. See my 1994, pt. I, § 11.)

A similar point applies to Maddy's (1997, pp. 146–52) discussion of the use of mathematics "outside of explicit idealizations." Her example is the use of the mathematics of continuous space-time. Maddy cites Feynman and Davies (and Einstein), who raise concerns that certain difficulties arise (*most likely*) because of the "assumption that space-time is continuous." She continues: "Under the circumstances, it seems the question of the continuity of space-time must be considered open" (p. 151).

It's hard to see, however, how this bears on the question of whether the *mathematics* so applied is true (or, for that matter, how it even bears on the ontological question of whether the items posited by such mathematics exist)—because all such physicists are hoping for a *successor* physical theory (1) in which such puzzles that plague current theories are solved and (2) that replaces the mathematics of continuous space-time with something else (something *better*). In other words, the mathematics of continuous space-time would, as Maddy explicitly notes, continue its role in an idealized (but false) empirical theory that we would find useful in certain (many) circumstances. But if this is right, then we've been given no more reason to take continuous space-time *mathematics* as false than we've been given to think false *any* branch of mathematics that facilitates the application of an instrumentally construed empirical theory.

Summary and Concluding Remarks

The concern in this chapter and chapter 1 has been with the Quine-Putnam indispensability thesis, specifically, with that version of it that argues that doctrine indispensable to scientific application must be regarded by us as true. This leads naturally to the question of when the truth predicate (or something logically powerful enough to replicate its powers) is ineliminable.

Along the way, the Quinean claim, that an indispensable empirical theory *must be* regarded as true, has been *qualified*, because such isn't the case in ordinary scientific practice if certain implications of such theories are placed in quarantine. Indeed, *idealizations* in science are typically the applications of theories that we take to be false in certain ways (and it's precisely these false implications that we quarantine).

Nevertheless, I've argued that this qualification offers little solace to instrumentally inclined philosophers—those, anyway, who urge us to take applied mathematical doctrine or fundamental scientific laws to be false—because in these cases scientists do *not* engage in the practice of placing in quarantine (some of) the implications of such theories, and so philosophers have no grounds to regard such theories as false, or even grounds to be agnostic regarding the truth of such theories.

I should say one last thing about the role of the truth idiom in this argument: Because the need for that idiom is, broadly speaking, pragmatic, one can certainly consider ways of avoiding truth ascriptions altogether. Let's say that such were possible, that the *global elimination* of the truth idiom from our discourse were successful, perhaps via, among other things, widescale augmentation of ontology—in Quine's sense—as contemplated in chapter 1. That is, suppose that blind truth ascriptions were unnecessary, that all applied doctrine involved finitely axiomatizable theories all of which we were able to write down; suppose that problems of convenience (the manageable exhibition condition) were also circumvented. Still, a *version* of our commitment to the truth of applied mathematical doctrine and to the truth of empirical law would remain in the form of the manifestation condition: No way of indicating a distinction between sentences we assent to and ones we treat instrumentally would be possible without the manifestation of that distinction in the inferential roles of these sorts of sentences within our body of beliefs. These two ways of arguing for the indistinguishability between classes of sentences in our body of beliefs are connected to each other by the Tarski-biconditional condition on the truth predicate: Assertion of a sentence on the one hand is materially connected to the assertion of the truth of that sentence on the other. (This identification, of course, exists only when a truth predicate is part of the vocabulary.)

A word about the status of the truth of unapplied mathematics: I've previously (in my 1994) argued that *all* mathematical doctrine, applied or unapplied, can be seen as true. This is the case even if mathematical systems are considered that contain prima facie incompatible axioms (e.g., the axiom of choice vs. the axiom of determinacy), simply because such systems can be isolated from one another by segregating them within separate languages or at least using different terminology (e.g., "set" and "set*," where sets obey the axiom of choice and sets* obey the axiom of determinacy). The point is the lightweight nature of "deflationist" truth—that nothing in the logic of truth prevents such a construal of mathematical doctrine. These considerations, however, don't force anyone away from an instrumentalist view of unapplied mathematics, a view, anyway, that takes unapplied mathematics to be neither true nor false, or as true and false as those values would be distributed over sentences that are actually about nothing at all. Fictionalism, if restricted to *unapplied* mathematics, is a view that I can't rule out on philosophical grounds (although I don't see much point to it, even in this case). *Applied* mathematics, however, is different: What I have shown in the foregoing is that the use of the truth idiom in science *prevents* instrumentalist construals of applied mathematical doctrine.

3

Criteria for the Ontological Commitments of Discourse

The classical version of the Quine-Putnam indispensability thesis involves two steps. The first step, which is the one opponents of the thesis have universally attacked, is to claim that indispensable application of mathematical doctrine requires our taking the mathematics so applied to be true. As I showed in chapters 1 and 2, it's hopeless to try to undermine the thesis by denying this claim.

The indispensability thesis, however, is that the indispensability of applied mathematical doctrine commits us to the *existence* of mathematical entities, and so we still need a path from the taking true of mathematical doctrine to a commitment to entities. This requires a method for identifying what entities we are committed to by espousing a doctrine—the application, as it's commonly put, of a *criterion* for what a discourse commits us to. The criterion of choice (almost universally) is Quine's (1948).

Here's the plot for chapter 3: I explore the reasons that Quine, and others, have given for his criterion, with a contrasting view implicitly in mind, that ontological commitment is to be carried not by the quantifiers but by an existence predicate instead. I will, in the process, go through a very long analysis of Quine's triviality thesis—this is because of the wide number of options that are possible defenses of it. I end this chapter on an indeterminate note: There is no good argument for Quine's criterion for what a discourse commits us to, at least when one restricts oneself to considerations about discourse. Chapter 4, therefore, naturally turns to other ways of determining a suitable criterion for what a discourse commits us to via the more metaphysical-looking route of a criterion for what exists.

Let's begin. The version of Quine's criterion I like is this: First, any body of doctrine D must be regimented into an interpreted first-order language D'.[1] That done, we take commitment to be recognized by implications of the form $(\exists x)\mathbf{S}x$, where $\mathbf{S}x$ is any formula with variable x free. That is, if $(\exists x)\mathbf{S}x$ is deducible from D', then D' commits its believers to $\mathbf{S}$s.

As Quine's (1948) discussion makes clear, this criterion is supposed to apply to discourses *regardless* of whether we take them as true. To the extent, therefore, that interpretation or translation of a discourse *requires* agreement, this criterion is designed to apply even to discourses that we don't understand—provided, of course, that we take them to be couched in first-order languages. A syntactic criterion suits this demand perfectly.[2]

Very oddly, philosophers often retain Quine's criterion while quite openly abandoning the restriction to first-order theories. I worry about the cogency of this (as does Quine) because the criterion, when restricted to first-order theories, really does offer a means of comparison between theories. When, however, one language has modal resources, for example, and another doesn't, it's unclear why Quine's criterion (alone) would allow straightforward ontological comparisons. I won't press this point further, however, because I deny that Quine's criterion can do what it's supposed to do, even in the restricted situations that it was crafted for.

---

[1] This is nontrivial, and there may not be consensus on *which* regimentation of a body of doctrine best reflects it. My objections to there being a definitive criterion for what a discourse commits us to won't turn on this issue.

[2] Restricted, as we are, to first-order languages, the criterion may be indifferently phrased deductively or implicationally. Of course, even if we use a model-theoretic notion of implication, the criterion still doesn't require an understanding or translation of nonlogical vocabulary in order to apply it to (foreign) discourse.

That Quine's criterion should be so glossed has been suggested before. Scheffler and Chomsky 1958–59 deplore this version of Quine's criterion on the grounds of its triviality (they write: "Not a very exciting result" (p. 81))—an objection Quine is unlikely to have recognized as an *objection*. They also claim (p. 82) that this version of the criterion serves no philosophical purposes "that cannot as economically be served by reference to the theories themselves." Here I demur. As economically? That's not obvious.

Cartwright 1954 considers several versions of the criterion but never quite suggests this one, perhaps because he takes ontological commitment to be a *relation* between sentences and collections of entities. Admittedly, Quine's own gloss on his criterion ("to be is to be the value of a variable") makes his text seem hospitable to the relation view of his criterion, but it's better to see that slogan as indicating the nature of the connection between objectual quantifiers and the semantics they're given in a metalanguage, especially because relational views, as both Cartwright 1954 and Scheffler and Chomsky 1958–59 make clear, don't even have a chance of working unless murky intensionalist maneuvers are introduced.

I like Gottlieb's (1980, chap. 2) defense of the version of the criterion I adopt, apart from two reservations: It's clear that Quine means his criterion to apply *only* to first-order languages, where the quantifier is standardly objectual, and it's also clear that he doesn't mean it to apply to individual sentences—except parasitically—but to whole *discourses*, which presuppose the use of first-order logic.

Here is a way of seeing how Quine might have attempted to establish his criterion. The first step (a momentous one for a philosopher deeply familiar with, and respectful of, Carnap's views on these matters) is to assume that ontological commitment genuinely takes place in ordinary discourse and is so understood to take place by ordinary speakers. The second step is to search through the terminological resources of first-order languages to determine what "parts of speech" of *regimented* discourse should carry ontological commitment.

It's important to remember that the regimenter is *not* out to recognize the *actual* logical form of ordinary language sentences. Regimentation is the *replacement* of ordinary language vehicles with something artifactual that serves certain purposes *better*. The regimenter, therefore, has no objection to the scientific study of natural languages by professional linguists, in particular, their study of the *actual* logical forms, if any, of natural-language sentences. That (descriptive) science is fully compatible with the beliefs (1) that nevertheless, natural languages are treacherous on the question of what the implications of our (scientific) theories are or what they (ontologically speaking) commit us to; and (2) that, in addition, we should take the ontological commitments and logical implications found within regimented languages (that we translate ordinary-language discourse into) as *normatively* constraining our claims about the *real* implications (and ontological commitments) of theories—couched in ordinary language—that we hold. (See Quine 1953a, pp. 149–50; 1960, pp. 157–61.)[3]

One can easily wonder why it isn't an entirely arbitrary matter which items in first-order languages are chosen to carry ontological commitment. After all, we don't *speak* such languages in the cradle (or anywhere *else* for that matter): They're constructions of mature (philosophical) science. This is correct, but in choosing an interpreted first-order regimentation of ordinary language, we must look to ordinary language and what's expressed there. That is, although regimentations are *self-interpreting*, the interpretations of their vocabularies operate with an eye to ordinary vernacular. This explains why we can't simply *stipulate* what items in a first-order language are to carry ontological commitment. We must recognize where in ordinary speech ontological commitment arises, and then we must examine our most natural regimentations of ordinary language to

---

[3] I should add that my take on regimentation (with its recognition of, and partial respect for, the presence of ontologizing in ordinary language) does not square well with *everything* Quine says on the subject. E.g., he says (1981a): "Ontological concern is not a correction of a lay thought and practice; it is foreign to the lay culture, though an outgrowth of it" (p. 9). My exegetical problem here is that I don't quite see how to get this more extreme thought—even with careful attention to its context—to fit compatibly with Quine's subscription to his triviality thesis regarding ontological commitment. So my regimenter has a more moderate view, one compatible with *much* of what Quine has written on regimentation and on explication. See the discussion later of Quine's triviality thesis.

see which idioms (conveying ontological commitment) are carried to which regimented items.[4]

What are the first-order candidates for bearing ontological commitment? Well, individual constants are certainly one natural candidate because ordinary names are regimented as such constants. Quine, applying Russell's theory of descriptions to first-order constants, transfers whatever ontological commitments ordinary-language names convey to the first-order quantifiers.[5] This move on Quine's part, notice, doesn't concern itself with how ontological commitment actually operates in ordinary languages: it's a technical point about the first-order items that ordinary names are regimented as: Those items (individual constants) can be defined away via Russell's theory of descriptions, and whatever ontological commitments they have (if any) become those of the quantifiers.

A different argument, which *does* turn on how names operate in the vernacular, may be found in Quine 1969: "'Pegasus', which is inflexibly a proper name grammatically speaking, can be used by persons who deny existence of its object. It is even used in denying that existence" (p. 93). Names, in the vernacular, don't *always* convey ontological commitments.

What about predicates? Quine (1948) deplores the recommendation that ordinary predicates convey ontological commitment; that is, he denies that an assertion of "Roses are red" should carry a commitment to a denotation for the predicate "red." But Quine never considers having a special predicate—call it an "existence predicate"—convey ontological commitment (*instead* of the quantifiers doing so).[6] As a result, he finds

---

[4] Of course, and this is one point of regimentation (as I've already stressed), we needn't follow where ordinary language dictates: We can legislate *changes* in the idioms of ontological commitment. See chapter 4.

[5] Some may worry whether Quinewise assimilation of first-order individual constants to assertions of existence and uniqueness conditions for descriptions enable the latter to serve as appropriate regimentations for ordinary-language names. One issue arises from Kripkean counterexamples to the view that names in the vernacular function like disguised descriptions. Quine, however, can accommodate these concerns (whether he's sympathetic or not) by using primitive predicates for the regimentation of vernacular names. Another problem, often recognized in the literature, is that the description approach doesn't give the right truth values to sentences such as "Sherlock Holmes is a fictional detective admired by many real detectives." I touch on this shortly.

[6] The existence-predicate suggestion is that an implication $(\exists x)\mathbf{S}x$ would not indicate ontological commitment to $\mathbf{S}$s. Only an implication $(\exists x)(\mathbf{S}x \ \& \ \mathbf{E}x)$, where $\mathbf{E}$ is the "existence" predicate, would indicate such commitment. See my 1997a, pp. 206–7.

Quine never explicitly attacks this idea. (I suspect because he thinks objectual quantifiers, on technical grounds alone, must be ontologically committing in any case. I discuss this shortly.) He does argue (1948, p. 3) against the idea of several grades of ontological status, *existence* and *subsistence*, say, but that's a different matter.

Notice that, given that the first-order quantifier carries ontological commitment, an existence predicate $\mathbf{E}(x)$ can be introduced like so: $(\exists y)(y = x)$; such a predicate would apply to everything in the domain of discourse—but that's *not* the idea of the first paragraph of this footnote. Rather (and this is a first, tempting but misleading, stab at the position because it occurs in a context, like this one, where phrases of the vernacular such as "there is"

the quantifier to be the sole remaining first-order candidate-indicator of ontological commitment.

Quine has subsequently endeavored to bolster the status of his criterion against objections by calling it trivial.[7] It should surprise the reader that Quine, the *regimentalist*, calls his criterion *trivial*, because this sounds like the choice of a criterion for ontological commitment for regimented languages isn't a matter of *legislation* (with an eye to the vernacular), whereas, of course, it has to be. Quine (1986) defends the triviality of his criterion this way:

> What there are, according to a given theory in standard form, are all and only the objects that the variables of quantification are meant in that theory to take as values. This is scarcely contestable, since "(x)" and "$(\exists x)$" are explained by the words "each object x is such that" and "there is an object x such that". Some languages may have no clear equivalent of our existential phrase "there is", nor of our quantifiers; but surely there is no putting the two asunder. (p. 89)

The *argument* for the triviality of Quine's criterion, thus, seems to be this: The ordinary language idiom "there is" carries ontological commitment, and this idiom is (straightforwardly) regimented as the (objectual) first-order existential quantifier.[8]

It's a *substantial* claim to assert that the ordinary language idiom "there is" conveys ontological commitment, and as I'll show, it's a claim that's not at all easy to establish. But before taking that (considerable) issue up, it's worth observing that Quine doesn't offer (here) a different argument that he raises elsewhere, that others have raised, and that most simply assume. This argument is that, for reasons not having anything to

---

or "what's in" are most naturally read as ontologically committing), **E** ranges over a "subclass of what's in the domain." Should this be described as the claim that "existence is a special property that not everything shares"? No—this *substantial* way of putting the position is misleading because (in this context) it makes it sound like there *are* things (which subsist?) but that don't exist; further these (subsisting?) things have properties—e.g., they don't exist. The best way to put the point (in this context) is from a perspective of semantic ascent: Our (regimented) terminology allows a statement "$(\exists x)\mathbf{P}x$" to be true even if nothing is a **P**. If, however, "$(\exists x)(\mathbf{P}x \ \& \ \mathbf{E}x)$" is true, then something is a **P**. (**P**s exist.) Does this mean there are things that subsist but don't exist in the *domain of discourse*? No—to put the point this way is really misleading. I say more about this shortly.

[7] Quine 1981c:

> Kripke writes congenially on ontology and referential quantification, stressing that their connection is trivially assured by the very explanation of referential quantification. The solemnity of my terms "ontological commitment" and "ontological criterion" has led my readers to suppose that there is more afoot than meets the eye, despite my protests. For all its triviality the connection had desperately needed stressing because of [certain] philosophers. (p. 174)

Also see Quine 1969, p. 97.

[8] See n. 3.

do with the vernacular at all, *objectual quantifiers are always ontologically committing.*[9]

It's easy to see why this looks true: Consider the semantics for objectual quantifiers. Informally, that semantics is presented this way: A domain of discourse must be given over which these quantifiers are to range, and that domain must be a collection of *objects*. The truth of an interpreted statement $(\exists x)\mathbf{S}x$ is then defined (nearly enough) as the satisfaction of the formula $\mathbf{S}x$ by *something* in that domain. How can objectual quantifiers, therefore, *not* be understood as ontologically committing?

This line of reasoning contains the crucial and unnoticed presupposition that the language in which the semantics for the objectual quantifiers is couched (the "metalanguage") is *itself* one with quantifiers that themselves carry ontological commitment. That is, the above considerations *force* us to regard our original set of (objectual) quantifiers as ontologically committing *only if* we regard the quantifiers in the metalanguage as ontologically committing. If we *don't*, then (metalanguage) *talk* of objects in a domain no more commits us to anything real than does the original claim, $(\exists x)\mathbf{S}x$, and this despite the original quantifiers still being uncontroversially objectual quantifiers in the technical sense.

My claim may look really confused. After all, presume the object language to have a "domain of discourse" (take it to be an "*interpreted* language").[10] When we provide a Tarski-style theory of truth for that object language, what's required, among other things, is that enough set theory be present in the metalanguage so that the domain of the metalanguage contains, among other things, ordered couples of objects from the domain of the object language.[11] How can any of this take place if the domain of the object language contains *nothing*?

---

[9] It's striking that both Barcan-Marcus (1971) and Parsons (1980) assume that use of objectual quantification convicts one of ontological commitments (to the domain of discourse). Parsons even assumes that his use of objectual quantifiers for nonexistent objects commits him to *nonexistent* objects. Also see Smart 1999, pp. 109–10, and Davidson 1999, p. 123, an exchange in which both sides virtually take it to be a *truism* that Tarskian satisfaction involves commitment to entities; and Schiffer 1996 (pp. 151-2, 156), where it's presumed that objectual quantification must bear ontological commitments (even if such commitments can be to kinds of "deflated" objects such as properties and fictional beings). One place I know of in the literature where it has been suggested that "objectual quantifiers" can be taken to be ontologically neutral (besides my 1997a and 1998) is in Landesman (1975). However, Routley (1980) may also claim this (I've not—as of this date—worked my way through the entire book). See, however, the seven paragraphs preceding the section "Some Important Illustrations."

[10] Why the scare quotes? I don't want the reader to *presume* that a "domain of discourse" or a language being "interpreted" amounts to (by *definition*) its terms picking out what exists. All that's meant is that such a language *isn't* a "meaningless formalism": It has a *subject matter*.

[11] This enables the definition of satisfaction for two-place relations: "$\mathbf{R}xy$" is satisfied by $\langle o_1, o_2 \rangle$ if and only if $\mathbf{R}^* o_1 o_2$ (where "$\mathbf{R}^*$" is a metalanguage relation coextensive with the object language relation "$\mathbf{R}$").

One should avoid bewitchery by images of object-language domains. When a metalanguage provides a Tarski-style theory of truth for an object language, it doesn't do so by nakedly handling the domain of discourse, reaching over to the object-language quantifiers, and placing that domain in their tight grip; no, rather, a body of sentences *are asserted* that provide the needed interpretation. For the predicates of the object language, too, it isn't that their extensions are just handed to them by the metalanguage; rather, it's that metalanguage predicates with the same extensions are used in the truth clauses.[12] And when couples of objects, which the object language relates to each other, are items that the metalanguage needs to talk about, existence claims (from set theory) are there to provide them.[13]

The lesson should be clear: Objectual quantifiers have "objects" to range over only relative to a body of claims in a metalanguage that itself gains access to those "objects," if at all, via what its own quantifiers range over. And if *those* (metalanguage) quantifiers do *not* carry ontological commitment, then neither do the objectual quantifiers that the metalanguage quantifiers help provide objectual semantics for. A slogan: *One can't read ontological commitments from semantic conditions unless one has already smuggled into those semantic conditions the ontology one would like to read off.*[14]

As I've mentioned, Landesman (1975) also suggests that "objectual quantification" is ontologically neutral. But his way of establishing his claim puzzles me. He seems to think that there can be objects to which we're committed ontologically in no sense at all (they don't exist, they don't subsist, etc.) and that nevertheless have properties that determine what's true or false of them. I can't see how to say in one breath that there are *objects* to which we refer and that have specific properties, and yet to which nevertheless we haven't incurred any ontological debts (of *any* sort). *What* exactly is *it* that we're supposed to be attributing properties *to*?

Were Landesman's a lonely voice in the wilderness, I might be tempted to stop the discussion of this position at this point. However, pretty much the same claim is made by the able logician and metaphysician

---

[12] How are such coextensional predicates "provided"? Well, it's *asserted* in the metalanguage that they have the same extensions as the object-language predicates they're paired with. What else is possible?

[13] I.e., a theorem of set theory such as $(s)(t)(\exists y)(x)(x \,\varepsilon\, y \Leftrightarrow x = s \vee^* x = t)$, and then a definition of ordered pairs using the sets guaranteed by that theorem, e.g., $\langle x, y \rangle =_{\text{def}} \{\{x\}, \{x, y\}\}$. See any standard book on basic set theory, e.g., Levy 1979, pp. 20, 24.

"Existence claims . . . are there to provide them": On the view *I'm* pushing, these aren't *genuine* existence claims. But mathematicians use such phrases as "existence proof" all the time. Are they misusing the word "exists"? Has the term been co-opted by them for a technical use (e.g., "ring," "field")? See chapter 4, where I give an answer to this question.

[14] The last six paragraphs explicate an argument from my 1997a, pp. 206–7, n. 19. It appears again (also in a fairly encapsulated form) in my 1998, p. 3.

Richard Routley (1980).[15] Routley's noneism is committed to the following principles (among others):

> M2. Very many objects do not exist; and in many cases they do not exist in any way at all, or have any form of being whatsoever.
>
> M3. Non-existent objects are constituted in one way or another, and have more or less determinate natures, and thus they have properties. In fact they have properties of a range of sorts, sometimes quite ordinary properties, e.g. the oft-quoted golden mountain is golden. (p. 2)[16]

Again, I *want* to describe this as incoherent. I don't believe you can claim that something exists in no sense at all—that it has no being (of any sort)—but then add (in the same breath, as it were) that *it* has specific properties and a nature that can be talked about. But maybe I'm wrong.[17] In any case, whether I'm right about this or not, the important point to make is that when *I* claim that objectual quantifiers are ontologically neutral, I'm *really* claiming that what falls outside the range of the existence predicate exists in *no sense at all*. Routley (see n. 16) makes a similar-sounding pronouncement, but there is a difference between his position and mine, and that difference *doesn't turn* on a tug of war over the phrase "in no sense at all," because Routley (and Landesman) clearly claim that such objects ("items") *have properties*, and I don't. And by "properties" they don't mean the mere truthful concatanation of predicates with singular terms that pick out *nothing*. When *I* describe how talk involving the nonexistent comes up, I speak of *stipulation* and *making it up*. There's nothing to *discover* about the properties of the nonexistent— there are no general (metaphysical) claims that can be made about the nonexistent except that they don't have any properties, and don't exist. Routley, on the other hand (M3) thinks there's something to argue for here, metaphysically speaking, and not merely something about word usage that indicates the presence of stipulation.[18]

---

[15] Both Landesman and Routley officially draw their inspiration for this point from Meinong. Nevertheless, two philosophers from such different (philosophical) backgrounds claiming the same (weird-sounding) thing? Then it *has* to be addressed (however bizarre one thinks it is).

[16] Here are some other illustrative quotes from Routley (1980): "Meinong maintained that although the number two does not exist it has being. On the new theory of number two neither exists nor is assigned being of any sort; however it does have properties, it has indeed a nature," and a few lines later, "the number two has no being of any kind; even so it is an object and can be talked about" (p. 4). Finally, we find that "'[i]tem' is introduced as an ontologically neutral term: *it is intended to carry no ontological, existential, or referential commitment whatsoever*" (p. 5).

[17] Maybe, that is, ordinary uses of "exist" (and "subsist," surely, because the latter is largely philosophical language untouched by ordinary usage) don't compel the claim that if something has properties and is not nothing then that's tantamount to claiming that it exists *in some sense*. I think, *ahem*, there's something to this. See chapter 5.

[18] Compare my discussion below of Quine's challenge regarding the individuation conditions for what doesn't exist with Routley's to be found in his 1980, pp. 411–26. On my view, the individuation conditions are simply made up (which is no surprise given that, on

The reader may fear that if things (e.g., Mickey Mouse) exist *in no sense at all* (in *my* sense of "no sense at all"), then nothing true can be said when "Mickey Mouse" appears as a noun phrase (as in, e.g., "Mickey Mouse doesn't exist," or "Many people think Mickey Mouse is cute"). This simply *isn't* true, and the insight that it isn't true is one that can be gleaned from the substitutionalist view of quantification.

For the insight that substitutionalists have, when they deny that their form of quantification need incur ontological debts, is that the sentences "Mickey Mouse is a fictional mouse" and "Mickey Mouse is a fictional dog," for example, can be taken to be, respectively, true and false without their having any *truthmakers* (to use currently fashionable language), without there being any *objects* the properties of which *determine* the truth-value status of those sentences. That is, the truth values of sentences containing terms like "Mickey Mouse" can be determined on the basis of factors other than the properties of an item, Mickey. And this is a very good thing (on my view) because things that don't exist—I have to say it again—don't *have* properties.

It's also this insight, by the way, that I'm taking advantage of when I claim that objectual quantification doesn't necessarily betray ontological commitments of any sort because objectual quantification is only a *body of semantic constraints of a certain sort couched in metalanguage sentences*, for we have to evaluate how those semantic claims *themselves* are made true before we can convict the object-language quantifiers of carrying ontological commitment.

Some Important Illustrations

It *can't* be understated how influential the view that first-order objectual quantifiers *must* carry ontological commitment has been. Because of this assumption, the analysis of the logical form of sentences, that is, the presentation of such sentences in a way that reveals why they have the implicational roles they have, is indissolubly wedded to *ontological revelation*. The exposure of logical form may be taken, as Quine takes it, as the regimentation of sentences from the vernacular or, as Davidson and others understand it, as the scientific study of the (actual) semantic structure of sentences in the vernacular. On the latter view, hidden quantifiers are often postulated as operating in such sentences to capture inferences to and from them. Although the need for such quantifiers is motivated by purely anaphoric needs (needs, i.e., for items that successive series of implications can rely on), when objectual quantifiers are imported to do this job, because they're presupposed to require ontological commitments, the analysis of logical form dramatically flowers into speculative metaphys-

---

my view, these things exist in no sense at all). Routley's view is *very* different: Because such objects have "determinate" natures, Routley is required to engage in metaphysical reasoning to determine what these natures *are*.

ics. On the regimentation view, these entities are required in our best regimentation of the vernacular; on the scientific study of the logical form of the vernacular view, these entities are ongoing (inadvertent) commitments of ordinary speakers.

Two Examples: Davidson (1967, 1977) is very explicit that the analysis of the logical form of adverbial inferences offers metaphysical insights. Consider the inferences from

>   (1) Jones nicked his cheek while shaving with a razor in the bathroom on Saturday,

to

>   (2) Jones nicked his cheek in the bathroom,

to

>   (3) Jones nicked his cheek.

Davidson (1977) writes:

> It seems that some iterative device is at work.... The books on logic... analyse these sentences to require relations with varying numbers of places depending on the number of adverbial modifications, but this leads to the unacceptable conclusion that there is an infinite basic vocabulary, and it fails to explain the obvious inferences. By interpreting these sentences as being about events, we can solve the problems. Then we can say that "Jones nicked his cheek in the bathroom on Saturday" is true if and only if there exists an event that is a nicking of his cheek by Jones, *and* that event took place in the bathroom, *and* it took place on Saturday. (p. 212)

Davidson quickly concludes that an ontology of events is required: He relies on the presupposition that the objectual quantifiers so introduced to explain obvious inferences and to avoid an infinite basic vocabulary bring ontological commitments in their train. I can't now go into the question of whether it's plausible or not to instead treat events as fictional items introduced for the reasons Davidson explicitly gives; this will turn on other issues about events. My aim here is only to illustrate how ontology is taken to ride on considerations of logical form alone, and to register a complaint against the presupposition in this practice that first-order objectual quantification *automatically* secures ontological commitments.

As a second example, Jackson (1976) argues that mental objects must exist because there is no way to preserve the logical properties of statements about mental states while eliminating the apparent relations of persons to the objects in such states, where such relations seem to commit us to *both* persons and these mental objects. So, for example, I may have a painful itch and a (distinct) ticklish itch. Further, my painful itch may burn, and the burning sensation of it may be intermittent. Further, the burning sensation of that itch may be the same burning sensation as a painful scrape on my knee. These ways of distinguishing and comparing aspects of my sensations are obliterated if we assimilate sensations to whole states of the person (at a time) or to adverbially characterized properties of the person. Up to this point, the argument demands quantification over mental objects because of the anaphoric needs the descriptions of relations between such items raise, and because of the apparent nonidentity conditions among such items.[19] The presupposition, that objectual quantification must be ontologically committing, is then invoked to make the move from the points about the logical form of mental-state expressions to a commitment to mental objects. I again can't go into the question of whether it's plausible or not to treat mental objects (instead) as fictional items, as manners of speaking, introduced for just the reasons Jackson explicitly gives; the plausibility of such a view turns on other issues about such purported objects. My only aim (again) is to illustrate how ontology is taken to ride on sheer considerations of logical form, and to register a complaint against the presupposition in this practice that objectual quantification automatically carries ontological commitments. *End of illustrations.*

A concern the reader may still have about ontologically neutral objectual quantification involves identity conditions. Such a reader remembers what Quine (1948) famously wrote of *possibilia*:

> How many of them are alike? Or would their being alike make them one? Are no *two* possible things alike? Is this the same as saying that it is impossible for two things to be alike? Or, finally, is the concept of identity simply inapplicable to unactualized possibles? But what sense can be found in talking of entities which cannot meaningfully be said to be identical with themselves and distinct from one another? (p. 4)

"No entity without identity," is the Quinean slogan, and in Quine's hands, this challenges the idea that objectual quantifiers, interpreted standardly, can nevertheless commit us to nothing at all—because things that exist in no sense at all can't have individuation conditions associated with them (so one would think); and Quine treats substitutivity of identity as "the criterion" of referential position, a position, that is, susceptible to a variable bindable by an objectual quantifier.[20]

The neat thing about entities that exist in no sense at all is that the identity conditions governing terms "referring" to such items are as made

---

[19] As these two examples already indicate, our ways of carving up subject matters *often* require the entrance of objectual quantifiers onto the semantic stage in *many* areas, with the resulting yield of items we'd otherwise be loathe to regard ourselves as ontologically committed to. An especially clear case, which I describe later, is that of discourse about fiction: a form of discourse that I argue involves talk of items we take to exist in no sense whatsoever. It's nice to think that the idioms that come up there, and that are best seen as nonontologically committing uses of "there are," in the vernacular can be nicely regimented by a first-order quantifier that is similarly noncommitting.

[20] See, e.g., Quine 1960, pp. 142, 167; 1950, pp. 75–6; 1953b; 1975, p. 102. In raising these concerns, Quine often acknowledges Frege.

up as the items themselves. To anticipate the position I argue for in chapter 4, I take the posits of pure mathematics to exist in no sense at all. Individuation conditions for mathematical posits, however, are stipulated in a way that mathematicians find most convenient for the practice of mathematics. That is to say, mathematical terms in different algorithmic systems are taken to "co-refer," because in this way mathematicians are able to profitably embed mathematical systems within one another.

For example, PA is embedded within stronger systems that have as theorems (some) sentences that aren't theorems of PA itself; similarly, the real numbers are embedded within the complex numbers by identifying real numbers with a certain subclass of complex numbers. The latter "ontological" way of describing the embedding is shorthand for descriptions of systemic embeddings. It's generally recognized that such identifications are arbitrary at least in the sense that there is more than one ideal way to do it, and no way to decide among these. Identifications between mathematical systems and, consequently, "co-referential" stipulations of the singular terms (and general terms) in those systems are made in ways that maximize the provability of theorems (short of inconsistency, of course). But they're also arbitrary in the sense that there are no external ontological requirements on such identifications: There are no mathematical objects that are ontologically independent of mathematicians and that they need to make sure mathematical terms successfully "(co-)refer" to.

What, then, *are* the conditions on two terms (in two different algorithmic systems) for their "co-reference"? Well, in general, there are no conditions (anything goes—we can construct any sort of *family* of algorithmic systems we want). But in a context where classical logic is presupposed, and where identity is present (and axiomatized to allow substitution of identicals for identicals *salva veritate*), the substitution of such terms for each other must not result in inconsistencies.[21] As a result, one requirement on identifying "co-referring" mathematical terms across systems is that the result be conservative with respect to the theorems of the weaker systems. (See Azzouni 1994, pt. II, §§ 6–7.)

Although such identifications are arbitrary, they're fixed for the duration, for purposes of proving theorems within a family of algorithmic systems, and for purposes of applying a family of algorithmic systems empirically.

I also—to anticipate the discussion on fiction shortly to come—take fictional entities to exist in no sense at all.[22] So, too, as in mathematics,

---

[21] Why not? Not for any metaphysically *deep* reason. In classical contexts, inconsistency yields triviality—everything is provable—and the resulting family of algorithmic systems is useless.

[22] By "fictional entity," here and hereafter, I mean those items in fiction that are *entirely* fictional. Neither Bush, should one appear in fiction, is fictional in this sense.

identity conditions on terms are stipulated, but because the demands of fiction are less stringent than those of mathematics (and because the corresponding issue of empirical application is nonexistent), concerns with fixing specific identity conditions for purposes of discussion are also less stringent. We may, for certain purposes, reidentify characters within a narrative who have changed in the interim, sometimes drastically. This seems to be a matter left totally up to the author's discretion. Other fictional media, such as comics or cartoons, stipulate reidentifications of characters across different cartoons or comics; in the process, characters can also change drastically or can even, as a result, possess contradictory properties. And, finally, there are identifications of characters across novels and other genres that we accept for some purposes and deny for others. For example, Sherlock Holmes has appeared in film as well as in novels and short stories. As critics discuss his role in such works, they'll identify him as the same character or not, as purposes suit. Such insouciance is no surprise, given that he exists in no sense at all, *and* that fictional purposes aren't better served by designing identity conditions more carefully.

But if a vocabulary item designating identity appears in the language in which we discuss fiction or fictional characters (or if the same identity relation is taken to appear in every family of algorithmic systems, for that matter), then the adoption of standard quantification plus bivalence seems to force contradictory claims: This is identical with that (for *these* purposes) but this isn't identical with that (for *those* purposes). That's correct, and the right conclusion to draw, I think, is that, strictly speaking, no term expressing *the* identity relation appears in these languages. Rather, what we have are a number of very useful pseudo-identity predicates, and sometimes we use one, and sometimes another, depending on our verbal needs.[23] This, given that such objects exist in no sense at all, should be no surprise: why should there be *real* questions about which are identical to which? In any case, there is no technical reason why the absence of genuine identity conditions would cause problems with objectual quantification.[24]

One terminological loose end: "Refer," as it's often used, is ontologically committing—if a term refers, then it refers to something; so too, if two terms co-refer, then they refer to the same something. I'll acquiesce

---

[23] Resnik 1997, pp. 209–12, when addressing the same problems, opts to treat mathematical objects, although "mind independent" as "incomplete objects": There is no fact of the matter about certain identity claims that we can formulate about such objects.

[24] Note: A standardly interpreted first-order language need not have *identity* among its vocabulary items, and its domain can certainly have more than one equivalence relation defined on it. In a case where the objects in that domain exist in no sense at all, there need be no residual worries that there is some genuine notion of identity implicitly at work, viz., that equivalence relation that most finely individuates the "objects" in the domain of discourse.

in this usage even though, as it's well known, Tarski's theory of truth allows us to define a "theory of denotation."[25] For, given what I've said about the ontological neutrality of objectual quantifiers, it follows that such a theory of denotation can easily be produced for terms that refer to nothing at all.

Nevertheless, when terms that refer to nothing at all occur in a language with identity, and with consistent identity conditions that allow statements like $A = B$ to be true for *distinct* terms A and B, I'll describe such terms as *co-referring** and will also, in general, speak of a term A as referring* in this sense even should it (like "Mickey Mouse" or "1," as I eventually claim) actually refer to nothing at all. I understand *referring** and *co-referring** as more general notions that include their narrower cousins *referring* and *co-referring*.

It should now be clear that the mere presence of quantifiers in the logical analysis of an idiom, even if they're objectual ones, needn't bring ontological commitments in its train. Still untouched, of course, is the Quinean claim (which is independent of the position that objectual first-order quantifiers are ontologically committing by virtue of their semantics alone) that ordinary uses of "there is" in the vernacular are understood by speakers to carry ontological commitment.[26] Notice that if this claim could be shown to be true, it would yield the ontologically committing nature of the first-order objectual quantifier even though the claim, that objectual quantifiers are ontologically committing by virtue of their semantics, fails. One problem, as I've mentioned, with *this* take on the triviality thesis is that it puts a lot of weight on ontological commitment as it's indicated in the vernacular. However, given the failure of the attempt to extract ontological commitments from the semantics of the quantifiers, the Quinean *has no choice*.

I turn now to an evaluation of the claim that uses of "there is" in the vernacular carry ontological commitment. Consider this ordinary sentence:

(M) There are fictional mice that talk.

---

[25] See, e.g., Field 1972.

[26] An especially neat use of this latter claim (which is *not* connected to any claim about formal quantifiers) may be found in Lewis 1986, pp. 8–17. The argument, briefly, is that the use of "could" in many contexts in the vernacular isn't served well by regimentation in standard modal notation because its role seems to be the broadening of the scope of quantifiers within its range so that possible objects fall under their domains of discourse. The conclusion that "we'd better have other-worldly things to quantify over—" (p. 17) has the ontological force Lewis wants (support for his modal realism) because of the already in place assumption that vernacular use of quantifiers carries ontological commitment. Also see Colyvan (2001, pp. 117–8), where his *presupposition* that the vernacular use of "there is" carries ontological commitment induces him to the conclusion that statements like "there is an even prime" or "there are differences between Clinton and Bush" *can't* be obviously true because it can't be *obvious* that differences or numbers *exist*. (Notice that

There seem to be six options regarding the ontological interpretation of the use of "there is" in (M),[27] each discussed in turn.

(1) The Paraphrase Option

Not all uses of "there is" are ontologically committing. But the ones that speakers utter that are *not* ontologically committing are ones speakers recognize to be usages of laziness. Speakers know that usages of laziness don't mean the same thing as genuine usages of "there is" that *are* ontologically committing, and such speakers (consequently) have paraphrases (in the vernacular) available to them that eliminate—for example—apparent ontological commitments to fictional entities, and so they can avoid the use of noncommitting instances of "there is" altogether should listeners be in danger of being confused about the speaker's actual ontological commitments.

Paraphrase is the stock-in-trade of the regimentalist: To paraphrase the vernacular in first-order logic *is* to regiment, and in doing so, the regimentalist may disregard the elusive goal of meaning preservation, and even the crisper goal of preserving the "logical form" of statements in the vernacular, as long as the first-order rendering of the original item from the vernacular can be used (broadly speaking) to serve the same roles as the original was used for.[28]

But paraphrase in support of Quine's triviality thesis is a more delicate matter. For here it isn't paraphrase into first-order logic that is the goal, but paraphrase within the *vernacular itself*. The claim in need of support (recall) is that lazy uses of "there is"—*in the vernacular*—don't carry ontological commitments. Given this, it might seem that although the regimentalist can usually disregard meaning preservation, he can't do so here: A paraphrase that *doesn't* preserve meaning *won't* prove that the original statement does not carry ontological commitments.

I don't think that proponents of the triviality thesis who help themselves to paraphrase are as hamstrung as the foregoing considerations suggest, because ordinary speakers are willing to countenance paraphrases that *don't* preserve meanings; in part this is because "same meaning" is

---

this is a *modus tollens* on the Carnapian-flavored view that such (internal) statements *are* ontologically insignificant, in part, because they *are* obvious.) Van Inwagen 2000, p. 239, straightforwardly embraces the claim that the vernacular "there is" is ontologically committing.

[27] Of course, philosophers are clever people, and no doubt other options will emerge. (*Indeed*, I actually go on to consider *seven*. See the discussion of the fourth option.)

By the way, I'm offering options (except for one) that will support Quine's triviality thesis; in doing so, however, I'm (in many cases) *not* offering options that *Quine* could stomach. But that's OK—I think the triviality thesis false—and this is so even if non-Quinean resources are mounted in its defense.

[28] This is why the paraphrase proponent is *not* committed to any particular way of producing paraphrases. See chapter 2, n. 18.

somewhat intuitively obscure even to excellent practitioners of the vernacular, and in part because ordinary speakers—just like the regimentalist—are interested in the *role* of a sentence for purposes at hand and will accept paraphrases that satisfy those purposes even if they *don't* preserve meaning. A third reason, perhaps the one most valuable to paraphrase proponents, is that paraphrases in the vernacular are often understood *not* to be intended to preserve meaning because the original statement doesn't express what the speaker "really meant." The paraphrase is then offered precisely to avoid an unwanted implication that might otherwise be drawn on the basis of that original statement. The hope, therefore, is that the paraphrases of (M) to follow are in this spirit.

These considerations may not satisfy *everyone* that paraphrasing within the vernacular in support of the triviality thesis needn't respect meaning preservation. But because I aim to show that *none* of the even remotely acceptable paraphrases of (M) within the vernacular will help proponents of the triviality thesis, I'll allow those proponents fairly wide latitude in choice of paraphrase.

In what follows, when I suggest that a locution *A* can substitute as a paraphrase for a locution *B*, I describe *A* and *B* as "coming to the same thing" (in that context), where "coming to the same thing," needn't involve synonymy. In particular, because "coming to the same thing" is, in general, relativized to context, it needn't, in general, be symmetrical (as synonymy, for example, is required to be), although it *sometimes* is.

So consider paraphrases of (M) such as (i) "Some mice that talk don't exist," (ii) "There are no mice that talk," or (iii) "There are *depictions* of talking mice." The first two paraphrases are ones that proponents of the triviality thesis hope can substitute for (M) in cases where the adjective "fictional" is operating not with reference to one or another fictional medium, but solely as a way of alluding to the nonexistence of such mice. (Paraphrase proponents *may* offer (ii) as a paraphrase when fictional discourse is being alluded to—but this is unwise, as I show shortly.) The second paraphrase is to be used where the practice of one or another fictional medium *is* being referred to, and it's understood as taking depictions to be real. But because depictions can be depictions of *X*s, without there being any *X*s at all, even fictional ones, paraphrases of the form (iii) don't commit their users to *X*s even though they *are* committed to depictions *of X*s.[29]

(2) The Meinongian Option

All uses of "there is" *are* ontologically committing. Indeed, (M), which is true, commits us to fictional mice. Fictional mice, of course, aren't

---

[29] On this view, "depictions of talking mice" is not understood to admit quantification into what follows "of." So it may, perhaps, be written as "depictions-of-talking-mice" or "talking-mice depictions" to avoid the unwanted impression that quantification into the phrase is possible.

real mice, no more than purported convicts need be convicts or alleged dogmatists dogmatists. But fictional mice *are* something (or, at least, they're not *nothing*—recall the earlier discussion of Routley's noneism), for all that, just like purported convicts and alleged dogmatists, although they don't exist, aren't real (in contrast to purported convicts and alleged dogmatists), because the following sentence is true, too:

(N) Although fictional mice can talk, no real mouse can.

So, the Meinongian conclusion is that some things that *are* don't exist or aren't real. Such things are *nonexistent* objects, and we're committed to their being such by our saying that there are such things.

Parsons 1980 offers something like this position. Notice, however, that option (2) comes in two versions. The first is that fictional mice and real mice are both kinds of *mice* (some mice are real, and some aren't). The second is that fictional mice aren't *mice*. If I *had* to adopt option (2), I'd go for the second version. This would allow me to accept both of the following natural statements as true: "There are fictional mice that talk" and "There are no mice that talk." Just because fictional mice are depicted *as* mice doesn't mean they *have to be* mice. And a good thing, too![30]

Some versions of this position take fictional objects to exist or to be real but to be abstracta or something else (concrete entities in fictional worlds, say). My objection to this option, given later, doesn't turn on subtle details about how any of this goes or on niceties about vernacular uses of the word "exists." Also, on some views of this, as I've mentioned, fictional mice *subsist* or have *being*, and on other views (although they're not nothing, and they do have properties) they don't. Details about this won't matter, either, as far as my objection is concerned.

(3) The Substitutional Quantifier Option

Uses of "there is" aren't *always* ontologically committing. This is because, although such uses are always regimented by existential quantifiers, that quantifier can sometimes be a substitutional one that needn't incur ontological commitments. In particular, we're allowed to assert (M) because we can assert:

(O) Mickey Mouse is a fictional mouse that talks,

and (M) follows from (O) by existential generalization.[31]

So this view supports the triviality thesis by arguing that uses of "there is" *in the vernacular* are ambiguous. One use is ontologically com-

---

[30] This distinction between properties *had* and properties *depicted* may be only terminologically distinct from van Inwagen's (2000, pp. 245–6) separation of properties that fictional characters *have* from those that they *hold*. Also see Salmon 1998, where a similar distinction is made.

[31] Barcan-Marcus (1971) argues for the usefulness of substitutional quantification to regiment mythological and fictional discourse without incurring ontological commitments.

mitting, and this is the use that supports the triviality thesis. The other use isn't. The two uses are distinguished (in the vernacular) because we recognize that a noncommitting use must be inferentially connected to an instantiation with respect to a name (the way that (M) and (O) are so connected), which is referentially empty.

(4) The Metaphor/Pretend Option

Uses of "there is" are always ontologically committing when taken *literally*. However, there are pretend or metaphorical uses of it with respect to fiction, mythology, or hallucinations, and perhaps in other cases, as well, where we pretend "there are" things but don't intend our language literally. This option allows metaphorical construal of usages of "there is" even when literal paraphrase of what's meant is *not* available; it's distinct from the paraphrase option because that *requires* paraphrase.[32]

(5) The Cancellation Option

The adjective "fictional" plays a very special role in (M). "There is" has, as part of its meaning, an ontological commitment to what there is stated to be, as well as an anaphoric role. But the ontological aspect of its meaning is (pragmatically) *canceled* by use of the adjective "fictional."

This sort of "cancellation" view has been popular in the literature, largely because of Grice and his followers. I distinguish two possible positions here. The first takes the meaning of "there is" to be ontologically committing but allows a *pragmatic* disavowal of an intention of commitment. In this case, it seems to me that (M) is then to be understood *nonliterally*—and this case is therefore a version of the metaphor/pretend option. The other construal takes "fictional" to function so that "there

---

[32] The view, if it's to support Quine's triviality thesis, requires distinguishability in the vernacular between metaphorical and literal uses. Yablo (1998; 2000, pp. 304–5) argues that such isn't to be had, and he draws the conclusion that Quine's criterion for what a discourse commits us to thus fails because it's understood to apply only to literal uses of words. Although I'm sympathetic to attacks on Quine's criterion, I don't like this particular way of going about it. To some extent, Yablo's argument is *ad hominem*: It attacks that textually relaxed and easy-going Quine who seems to allow that certain usages are metaphorical even without the availability of paraphrase. I interpret Quine more severely on this point: Without paraphrase, there can be no claim of metaphorical usage *that allows us to repudiate commitments to what's metaphorically referred to*. This, by the way, makes Quine's particular brand of physicalism programmatic: Until paraphrases are available, if I must refer to cities or angry clouds to express what I want, then I'm *committed* to cities and angry clouds. Period.

We can weaken Quine's constraint, as I indicated in chapter 2, but not in a way that supports Yablo's attack on Quine's criterion: Sentences we take literally, and believe, are ones all of whose implications are part of our body of beliefs. Sentences understood metaphorically are ones not all of whose literal implications are part of our body of beliefs. This provides a way of distinguishing metaphorical from literal usage of sentences, even without a full explication of what that metaphoricality consists in.

is" (because of the cancellation) is *stripped* of its ontological implication. In this case, we've got something similar to the substitutional quantifier option, although *without* the use of substitutional quantification. The first version of this view is open to the objections against the metaphor/pretend option that I discuss later. The second version, however, is untouched by the objection to the substitutional quantifier option because that objection focuses on shortcomings with this application of substitutional quantification. I raise fresh objections to this version of the cancellation option.

(6) The Ontologically Neutral Anaphora Option

Generally, uses of "there is" *are not* ontologically committing. Its primary uses are, grammatically, to introduce subject matters and permit anaphora, or to keep track of certain inferences, regardless of whether what's spoken of exists in any sense at all. If what's spoken of exists in no sense at all, then statements that involve nondenoting descriptions and names are still distinguishably true or false ("Mickey Mouse is a fictional mouse": true; "Mickey Mouse is a fictional duck": false), but such statements aren't made true or false by "truthmakers," objects to which we're committed, and the properties of which determine the truth-value status of our statements.[33]

Any of options (1) – (5) can be used to support the triviality thesis. So let's take them in turn.

On the Paraphrase Option

The paraphrase option faces a really tough challenge, as many have noted. The problem is, as I'll illustrate, that although the second candidate paraphrase ("There are no mice that talk") does the job in cases where what's being claimed is simply the nonexistence of something, the third ("There are depictions of talking mice"), which is the only real option for statements about various sorts of fictional discourse, won't work.[34] Let's con-

---

[33] In my 1997a, p. 206, I distinguished two roles for the quantifiers: an inference-licensing role and an ontologically committing role. I now think "there is" and its kin in the vernacular only have inference-licensing and anaphora-supporting roles, and that ontological commitments, when made, are based on more elusive context-dependent indicators (see chapter 5 for further remarks on this). Hofweber (2000) also distinguishes two such roles for the quantifier but marks the inference-licensing role as one that arises when information about the subject matter is "incomplete." This looks wrong: Quantifiers are used when (but not necessarily only when) names are absent, *regardless* of how much is known about the subject matter.

[34] Of course, a strategy *Quine* might avail himself of with respect to discourse about fiction is similar to one he has been tempted by in the case of unapplied mathematics: treat the use of "there is" in and about fictional discourse as just not part of our web of beliefs—strictly construed. Then paraphrasing away "there is" in the vernacular becomes unnecessary. See the digression later for why this isn't a good move.

sider the first paraphrase, "Some mice that talk don't exist." The problem with this item is that the locution "some mice" as used here seems to come to the same thing as "there are mice," so "Some mice that talk don't exist" seems to come to the same thing as "There are mice that talk but that don't exist." But then, arguably, we've still got a use of the locution "there are," which isn't ontologically committing: "There is" isn't ontologically committing in contrast to the predicate "exists" that does carry commitment (at least in this sentence).

Still, the paraphrase proponent can try to avail himself of a legality at this point: The issue is whether "There are *M*s" is ontologically committing in the vernacular; it's not whether "Some *M*s" are ontologically committing. So the fact that we can paraphrase away "There are *M*s" this way tells us that "There are *M*s" is *ambiguous* and that the paraphrase proponent need only attach the success of the triviality thesis to one of those meanings. We have (1) an ontologically committing use and (2) an ontologically noncommitting use. The latter is recognized because it comes to the same thing as "Some *M*s." Unfortunately, this ploy doesn't succeed in cleanly separating two uses of "there is" in the vernacular because "Some *M*s" can be substituted for "There are *M*s" equally well regardless of whether "There are *M*s" is taken to be ontologically committing or not.

So let's now consider the paraphrase of (M) that characterizes it as "There are no mice that talk." This option works perfectly well when what's being paraphrased in the vernacular is a simple denial of the existence of something. But I should make it perfectly clear that it's unacceptable if what's involved is a claim about what goes on in fictional discourses of various sorts. In that case, there are not, presumably, fictional mice willing and able to do *anything at all*. There are no fictional mice, for example, that eat the planet Neptune in one gulp.[35] Also, there are many things fictional mice do that real mice also do (eat cheese, for example). But this paraphrase makes each statement about what fictional mice do *true* if real mice don't do what the statement has fictional mice doing, and false otherwise.

It may be thought that paraphrasing via depictions does better because it makes the implicit reference to a fictional medium explicit: "There are depictions of mice that talk" is true; "there are depictions of mice that eat the planet Neptune in one gulp" is false—just as it should be. But consider this sentence: "There are fictional detectives who are admired by many real detectives." Paraphrasing this via depictions can give sentences with the wrong truth values: "There are depictions of detectives that are admired by many real detectives" *may be* true, but one is perfectly capable of admiring the fictional detective depicted without

---

[35] I hope this is right: I hope no novel (play, cartoon, or *whatever*) exists in which a mouse eats the planet Neptune in one gulp. But if so, substitute some more outlandish example, or use a simple contradiction: "There are fictional mice that aren't fictional mice," or perhaps "There are fictional mice that aren't depicted as mice."

admiring his depiction (e.g., one may admire Sherlock Holmes but *utterly deplore* Doyle's prose style, or vice versa). This problem isn't easily surmountable by paraphrase: We think about, admire, and react to characters in fiction, mythology, and so on—this is how we *talk*, anyhow; and we do so even when entirely aware that such things don't exist in any sense at all. Nonetheless, these thoughts, admirations, and reactions aren't—or at least aren't *naturally*—construed as thoughts, admirations, and reactions to depictions or text or anything like that. This is what makes these paraphrases poor options.[36]

Some may want to fault this sort of counterexample by noting that it's an opaque context, and thus will counsel banning quantification into such a context. The idea is that the logic of propositional attitude terms is tangled enough that paraphrase proponents—even though they're restricted to the vernacular for their paraphrases—can ignore such purported counterexamples. This line, I think, is mistaken, not because propositional attitude contexts don't pose formidable logical problems, but because those problems have nothing to do with the issues at hand.

The *real* problem is that the depiction option seals off the content of what's depicted from quantification, and so anaphora becomes impossible *even when* the context isn't a propositional attitude one or, more generally, an opaque context. Consider this example:

> There is a fictional mouse who is (sometimes) depicted as having a round friendly face (and big eyes), and who is (sometimes) depicted as looking quite unpleasantly ratlike (with small eyes).

How do we paraphrase this? The problem is that we can't paraphrase it as the claim that there is a fictional mouse who is depicted with both a round friendly face (and big eyes) *and* with a ratlike small-eyed face. That might be true of some fictional Jekyll-and-Hyde mouse who changes faces (in the same cartoon) after drinking a potion, but it's not true of this one (Mickey, to name names). If we paraphrase the first part as about a Mickey-Mouse depiction, even a round-friendly-faced-with-big-eyes-Mickey-Mouse depiction, we can't refer back to the mouse so depicted and describe *him* as something that is sometimes depicted as having a ratlike face with small eyes. At best, we can do this: There are Mickey-

---

[36] Walton (1990, p. 203) denies this, and there has been quite a bit of discussion about whether he's right or wrong in so doing. I think he's wrong, but that needn't detain us now because, in any case, the paraphrase proponent won't find this strategy of much use. For suppose there is a kind of special "pretending*" that we engage in when we "have" emotions toward characters we know to be fictional: if we replace "There are fictional detectives who are admired by many real detectives" with "There are fictional detectives who are pretend*-admired by many real detectives," we still have the same paraphrase problem. Walton's *ultimate* solution—which I discuss later—is more drastic than anything the paraphrase proponent will want—it requires extending the pretense to the proposition *itself*: "There are fictional detectives who are admired by many real detectives" doesn't *really* express a proposition; we only *pretend* that it does.

Mouse depictions that are round-friendly-faced-with-big-eyes-depictions, and there are Mickey-Mouse depictions that are rat-faced-with-small-eyes depictions. Lest the reader think this suffices, notice that it can be that *Mickey Mouse* is repulsive in the second depiction but not in the first. This isn't captured adequately by "There are Mickey-Mouse depictions that are rat-faced-with-small-eyes and which are repulsive." (Again, the depiction might be quite pleasant even if the mouse depicted in it isn't.) What must be said instead is this: There are Mickey-Mouse depictions that are rat-faced-with-small-eyes-and-repulsive.

Notice how the issues raised in the section titled "Some Important Illustrations" are coming up here (which is no surprise, because the anaphoric motivation is the same in *all* these cases). Suppose we want to infer from the fact that there are depictions of Mickey Mouse in which *he* is rat-faced, has small eyes, and looks repulsive, to there being depictions of Mickey Mouse in which *he* is rat-faced and has small eyes, to there being depictions of Mickey Mouse in which *he* has small eyes. Try representing these inferences via adjectival constructions on depictions, and then recall the Davidson quote.

In addition, I want to stress that we're considering paraphrases in the vernacular available to speakers, not clever regimentations; the requirement is that such paraphrases be ones that ordinary speakers are likely to shift to in order to avoid being misunderstood. It seems clear that the depiction option is driving us to paraphrases that only a philosopher would think of. Ordinary speakers, it seems, do what's natural, refer* to Mickey Mouse, and describe *him* as repulsive when depicted as Disney first did, and as much more attractive (i.e., *Mickey Mouse* is much more attractive) as depicted later. Ultimately, this is why the paraphrase option—applied to the vernacular as it must be to defend Quine's triviality thesis—fails.

One last loose end: The reader may wonder if "depicts," contrary to my suggestion, is an opaque context. Can't something (a work of fiction) depict Tully without thereby depicting Cicero? *Possibly*, but we can always restrict ourselves to uses of "depict" that aren't opaque—that is, to uses that don't violate the identity conditions for the fictional terms already in place:[37] The problem arises anyway. This is relevant because Quine (early and late) regards propositional attitude ascriptions, in particular, as indispensable. This means that such opaque contexts must be logically tamed or, at least, regimented when possible by logically tame idioms, where "logically tame" means "extensional"—the substitution of co-referring expressions is truth-preserving. I've extended the requirement to co-referring* terms—but apart from that, the condition is the same one. I will sometimes describe the condition as "extensionality*." My point, therefore, is that even when propositional attitude ascriptions and talk of

---

[37] I.e., when fictional items are being discussed, and two distinct terms A and B are co-referential*, then a depiction is a depiction of A if and only if it is a depiction of B.

depictions are well behaved, so that the substitution of co-referring* terms is truth-preserving, problems with anaphora remain, and it's *those* problems that ultimately sink the paraphrase option.

On the Meinongian Option

It's worth reminding philosophers (who seem otherwise strangely prone to forget this) that fictional objects are paradigmatic example of items *we make up*, items that exist *in no sense whatsoever*. Russell (1919) eloquently puts the point this way:

> [M]any logicians have been driven to the conclusion that there are unreal objects. . . . In such theories, it seems to me, there is a failure of that feeling for reality which ought to be preserved even in the most abstract studies. Logic, I should maintain, must no more admit a unicorn than zoology can; for logic is concerned with the real world just as truly as zoology, though with its more abstract and general features. To say that unicorns have an existence in heraldry, or in literature, or in imagination, is a most pitiful and paltry evasion. What exists in heraldry is not an animal, made of flesh and blood, moving and breathing of its own initiative. What exists is a picture, or a description in words. Similarly, to maintain that Hamlet, for example, exists in his own world, namely in the world of Shakespeare's imagination, just as truly as (say) Napoleon existed in the ordinary world, is to say something deliberately confusing, or else confused to a degree which is scarcely credible. There is only one world, the "real" world: Shakespeare's imagination is part of it, and the thoughts that he had in writing *Hamlet* are real. So are the thoughts that we have in reading the play. But it is of the very essence of fiction that only the thoughts, feelings, etc., in Shakespeare and his readers are real, and that there is not, in addition to them, an objective Hamlet. When you have taken account of all the feelings roused by Napoleon in writers and readers of history, you have not touched the actual man; but in the case of Hamlet you have come to the end of him. If no one thought about Hamlet, there would be nothing left of him; if no one had thought about Napoleon, he would have soon seen to it that some one did. The sense of reality is vital in logic, and whoever juggles with it by pretending that Hamlet has another kind of reality is doing a disservice to thought. (pp. 169–70)

Nearly a century later, Deutsch (2000) disagrees. He thinks we're all "casual Meinongians." Why? Because "our talk about fictional things and that of authors, often takes the form of descriptions of things, and events. . . . So such talk invokes, perhaps naively, a Meinongian-like domain of discourse" (p. 164). We've seen this before, but it's worth repeating: A subject matter, however replete with descriptions, doesn't *suffice* for the existence (in *any* sense) of objects (of *any* sort), however informally (or casually) this is taken.

Salmon (1998) treats fictional singular terms as referring to ("man-made") abstracta, and he argues (pp. 293–5) that his view (as well as other views that take such terms to refer to abstracta, such as van Inwagen's and Kripke's) shouldn't be described as "Meinongian," because such views don't attribute referents to all manner of singular terms, such as "round square" or "current French monarch," and also that such views remain within the spirit of Russell's quote (which Salmon gives in full) because such abstracta aren't "lower class," ontologically speaking. I don't agree. It's robustly part of common sense that fictional objects don't exist *in any sense at all*, and this means they aren't abstracta *either*.[38]

Thus, *any* theory of discourse about fiction must, unless there are *very* good reasons not to, have built into it that fictional objects *are* in no sense at all. It's true that if (all) the alternatives to the Meinongian option have even less palatable consequences, then that option must be taken seriously, perhaps even adopted. Less palatable consequences, for example, are those of the paraphrase option: paraphrases that prevent anaphora and that don't preserve truth value. These are grave flaws for any construal of discourse about fictional entities. But a commitment to fictional objects is also a grave flaw, and I'll argue that the ontologically neutral anaphora option is better.[39]

On the Substitutional Quantifier Option

The substitutional construal of ontologically neutral uses of "there is" faces a fatal objection. This is that many uses of "there is" in discourse about fiction don't involve characters with *names*, and this blocks the use of substitutional semantics in those cases. The following sentence is true (as even a cursory glance at fairy tales shows):

(P) There are fictional animals that talk and that don't have names.

This shows that regimentations of noncommitting uses (if any) of "there is" require objectual semantics.[40]

---

[38] A sensible ordinary person will find it bizarre if you claim that fictional objects exist, either in the "strict" Meinongian sense or in the sense of being abstracta. Indeed, *many* sensible ordinary people will find it bizarre if you claim that abstracta exist—they will, anyway, once you've made it perfectly clear what you really *mean* by this.

[39] Salmon's motivation is clear: True (and false) sentences with nonreferring expressions pose a serious prima facie problem for his strong *Millian* view, "according to which the semantic contents of certain simple singular terms, including at least ordinary proper names and demonstratives, are simply their referents, so that a sentence containing a nonvacuous proper name expresses a singular proposition, in which the name's bearer occurs directly as a constituent," and even a (prima facie) problem for the "less committal" "*theory of direct reference*, according to which the semantic content of a name or demonstrative is not given by any definite description" (Salmon 1998, p. 278).

Nevertheless, a general theme of *this* book is that if a theory needs quantification over "entities" of a certain sort, it doesn't follow that such things exist. Although I can't get into details about proper names in *this* book, the moral is the same one for semantic theories as it is for any other scientific theory. I say a *little* more about this shortly; see esp. n. 55.

[40] I'm surprised not to have found this argument in the literature already, because many characters in fiction are nameless, e.g., *most* of the soldiers who fight in the Trojan War in *The Illiad*, or most of those who live and die in the background of *War and Peace*.

On the Metaphor/Pretend Option

By way of raising an objection to the metaphor/pretend option, consider the following sentence:[41]

> (Q) While working on the project, Susan caught fire although David didn't catch fire.

Although it's easy to provide contexts where both "caught fire" and "catch fire" in an utterance of (Q) can be understood metaphorically (as standing for "got really enthusiastic"), and more somber contexts where both in an utterance of (Q) can be understood literally, it's extremely hard, if not impossible, to provide a context where a single utterance of (Q) can be understood so that "caught fire" is metaphorical whereas "catch fire" is literal (or vice versa). Part of the problem is that no listener will understand an utterance of (Q) so meant. But part of the problem is that (Q) marks a contrast, and the contrast has no point unless "catching fire" and "caught fire" are both understood literally or both understood metaphorically. Now consider:

> (R) There are fictional mice that talk, although there are no real ones that talk.

This is an entirely natural sentence, and the naturalness of the contrast drawn can only be explained if both uses of "there are" are literal (or both metaphorical).[42]

Are both uses of "there are" in (R) metaphorical? It's hard to see how the second appearance of "there are" is: The clause it appears in looks like a straightforward assertion. Walton (1990, chaps. 10 and 11), however, offers a more drastic possibility. This is that speakers sometimes only *pretend* to assert propositions.[43] On this view, should someone utter

---

It's understood that there are many more hobbits, dwarves, and elves in Middle Earth than those that literally appear on the pages of *The Lord of the Rings*. By the way, *depicting* a character as having a name isn't the same thing as the discourse actually giving the character a name. (We can presume that all the characters who are understood to die in *War and Peace* are understood to have names, too.) Substitutional semantics for discourse about fiction can't manage with mere depictions of characters *as* having names. There are desperate maneuvers still available, of course. One can understand fictional discourse to be far more bloated than it appears to be (containing names for every character alluded to—even though neither author nor readers have any idea what these could be).

[41] Recall that this is the suggestion that there are metaphorical or pretend uses of "there is" in the vernacular that aren't ontologically committing.

[42] Yablo (2000) argues that "there is" and "exists" are metaphorically used in a wide range of cases. The contrast argument just given generalizes to these examples in a way that makes this claim extremely dubious, unless what's meant is that these locutions are *always* used metaphorically. I rule this latter suggestion out in what immediately follows.

Jeff McConnell has pointed out to me that a synonymous version of (R) is "There are fictional mice that talk, but no real ones that do." Thus, the two appearances of "there are" in (R) cannot differ in literality.

[43] Despite my label "metaphor/pretend option," I do *not* mean to suggest that the claim that "there is" is sometimes metaphorical is *identical* to the claim that we sometimes "pretend" to utter propositions with "there is" in them. Walton (2000) carefully distin-

"Mickey Mouse is a talking mouse," it's not that any particular word in that sentence is being used metaphorically; rather, it's that the presence of a fictional name reveals that the speaker is only pretending to refer (by means of "Mickey Mouse") and that (really) no proposition has been uttered at all, but only fictionally so, or in pretense. Fictional names are ones that speakers can *never* use in ordinary propositions but can only *pretend* to refer with.[44] Not only do we pretend to refer when we use fictional names, but also (Walton 1990, pp. 428–9) we pretend to attribute properties to things when we use predicates such as "mythical," "illusory," "imaginary," "fake," "real," "genuine," and so on. Thus, in uttering (R), we only pretend to mark a contrast between fictional mice and real ones.

I find this view too drastic in two respects. The first is that it generally requires discourse about fictional objects to be meaningless. That is, not only is fictional discourse itself meaningless on this view, but discourse on the part of literary critics *about* characters in fiction is meaningless, as well.[45] However, just as it's a grave drawback for a theory to commit us to fictional objects (because fictional objects are paradigmatic of what doesn't exist in any sense at all), it's also a grave drawback to require palpably meaningful discourse to be meaningless. On both these points, the paraphrase option does much better: It treats fictional discourse as meaningful and fictional objects as nonexistent (in every sense).

A second drawback of this view is that speakers are taken to be pretending to refer, and as a result, to be pretending to be uttering propositions, even if speakers (sincerely) disavow themselves as engaged in pretense. The problem isn't that speakers can't often sincerely disavow true claims about themselves: Speakers are notoriously ignorant about their own motivations, for example. But in this case an ordinary psychological state, pretending, has been drafted as a device that allows us to avoid the difficult task of analyzing the logical forms of sentences in such a way that their logical properties are preserved, in particular, what implications they seem to have and what truth values speakers attribute to them.[46] And this

---

guishes his pretend-doctrine from the claim that, e.g., "exists" is sometimes used metaphorically. The state of play is this: I regard the metaphoricality claim as refuted by the immediately preceding remarks on (R), and I'm allowing the proponent of the metaphor/pretend option to adopt a more extreme position toward (R), viz., some version or other of Walton's claim that (R) isn't (really) a proposition *at all*.

[44] Salmon (1998) attributes the view to Kripke that "a sentence using a proper name in an ordinary context (not within quotation marks, etc.) expresses a proposition only if the name refers" (p. 312, n. 19). Donnellan (1974) writes: "If a child says, 'Santa Claus will come tonight,' he cannot have spoken the truth, although, for various reasons, I think it better to say that he has not even expressed a proposition" (pp. 20–1).

[45] No doubt a great deal of literary criticism *is* meaningless. But surely it doesn't *have* to be.

[46] See Walton 1990, pp. 416–9. Walton *does* offer paraphrases, but such paraphrases "seek to capture ... what speakers say in uttering the sentences cited, not what the sentences themselves mean or what propositions *they* express, if any" (p. 417). They are thus not required to respect the logical form of the sentences they paraphrase, or even to be sentences speakers would recognize themselves as having *said*.

raises the most important objection to this approach, that it has insulated itself against the only data available to discriminate among alternative theories of this sort. If the claims of speakers to have asserted meaningful statements in a sincere way (without intent of pretense or play) can be denied for purely theoretical reasons, in what sense can the resulting theory be regarded as sensitive to evidence?[47]

A Digression

Semantic pretense views on large chunks of our discourse—fiction, discourse about fiction, mathematics, propositional attitude ascriptions, even truth attributions—are becoming fashionable, and I've detected gestures toward the previous objections, in one form or another, at conferences and in the literature. I find them unanswerable (at least insofar as they provide evaluative comparisons with the other candidates I consider), however, and so I've given them in my own way for the record. I should add that I *like* the view that discourse in *fiction itself*, in novels, short stories, and so on, involves pretense, provided the view is taken that the statements of fiction are *pretended-true* rather than the view being that a pretense is made of asserting statements *themselves*.[48] But I don't want to extend *this* view to the sorts of claims about fictional characters I've considered above (e.g., (M))—because these are statements not pretended-true but actually true; and for similar reasons, I'm loathe to extend the pretend-true view to mathematics or to truth attributions themselves.

This point is worth expanding on.[49] The literature on pretense has *not*, to my knowledge, engaged itself explicitly with the issue of the Quine-Putnam indispensability thesis—but it *has* to if its program is to work. The problem is that statements we're pretending to utter, on this view, must be ones that can be cleanly separated from ones that we're not pretending to utter, on pain of *global pretense*: that our *entire* body of utterances collapses into pretend utterances.

Certainly, when it comes to the pretense take on abstracta or on truth, this problem must be faced, because what fuels the indispensability argument right from the beginning is that (in the important cases) an

---

[47] A way of seeing the point about evidence is to ask, "Where will this all *stop?*" Why not, e.g., claim that we have no ontological commitments at all? No matter *what* we say, we're always pretending, even though we don't think we are and our statements seem to have logical forms that we respect when we draw inferences from them. Call this the threat of *global pretense*, and see the digression immediately below.

[48] This is *similar* to Urmson's (1976) view: "[M]aking up fiction is not a case of stating, or asserting, or propounding a proposition and includes no acts such as referring" (p. 155). On the view I like, however, "pretending-true" does *not* exclude statements *in* fiction—some of them, anyway—from being *actually* true, and it does not exclude certain names, "Nixon," for example, from referring.

[49] The pretend view, apart from its application to sentences *in* fiction itself, is open to the same objections I raised against the instrumentalist strategy (B3) in chapter 2, simply because it *is* an instrumentalist strategy, as I indicated there. Nevertheless, some additional remarks are in order.

empirical subject matter and the mathematics applied to it can't be cleanly separated as the nonglobal pretense-theorist needs. If one restricts the examples of application of mathematics to simple ones, for example, "The number of cats is the same as the number of dogs," where there is a sharp cleavage between mathematical and nonmathematical vocabulary, one can get the impression that the indispensability of applied mathematics is only due to the need to circumvent (the impossible) explicit presentations of certain sorts of infinitary statement. The view, then, amounts to the claim that one pretends to utter a statement quantifying over numbers but really means to indicate a certain set of sentences too large to otherwise express. Unfortunately, as I show in part II, this view presupposes a false picture of how mathematics is applied: Usually the subject matter to which mathematics is applied doesn't admit of a mathematics-free construal (not even if we allow ourselves the resources of infinitely many abstracta-free sentences).[50]

Can *some* chunks of our discourse be isolated enough to apply the pretense view to them *without*, as a result, tarring the rest of our discourse with pretense? What about (in particular) discourse about fiction? Surely Quine, gripped as he is by his vision of an austere science couched in the pure language of physics (plus set theory), may be so tempted to dismiss discourse about fiction this way. Alas, he should not do so: Even without eliminable paraphrase into his austere vocabulary, those parts of our body of beliefs, even remarks *about fiction*, are subject to confirmation holism, the methodological constraint that evidence for *anything* can come from *anywhere*.[51] No part of our body of beliefs, therefore, should be isolated

---

[50] A *confession*: I've mistakenly implied (Azzouni 1997b, pp. 480–3) that *all* applications of mathematics are solely a matter of needing to expand our vocabulary in certain respects. So, e.g., if we are unaware of the upper bound on the number of cats and dogs (and we want to use first-order quantification, but avoid quantification over numbers), we need an infinite-length sentence to express the claim that the number of dogs is the same as the number of cats. Similarly, in order to describe the locations and distances of objects in space to an arbitrary degree of fineness, an open-ended number of "place-predicates" must be replaced by quantification over points. Yablo (e.g., 2000, pp. 295–6) is prone to give similar examples. It's no surprise that one thinks of such illustrations as paradigmatic applied mathematics, because they've been staple fare in the literature for a long time, both for those who find ontological commitments thus required unproblematical and for those who deplore them utterly (e.g., Quine 1972a, p. 242; Gottlieb 1980, p. 42). Unfortunately, these examples are very misleading precisely because they only involve the application of mathematics to an empirical vocabulary that is mathematics-free, and so they give the impression that a kind of logicism is within sight if we only treat all applications of mathematics (as these examples suggest we can) as matters of metaphorical or pretense expressions of commitment to mathematical abstracta the literal content of which is the truth of a certain (infinite) class of sentences all the members of which are actually mathematics-free in their vocabulary. As I show in part II, most applications of mathematics to the sciences are *not* amenable to this treatment.

[51] In particular, evidence for theories couched in austere vocabulary, e.g., physics, can come from phenomena only describable in nonaustere vocabulary. Indeed, evidence leading to quantum mechanics came from (unreduced) *chemistry*. Colyvan (2001, p. 111, n. 30) mentions another example (neat—but not one, it seems to me, where there was actually

unless there are *epistemic* reasons to do so (as discussed in chapter 2 in considering the instrumentalist strategy (B3)). *End of digression.*

On the Cancellation Option

The cancellation option[52] supports the proponent of the triviality thesis because, on either of the two possible ways of taking this approach, (1) every usage of "there is" is still ontologically committing (and we just pragmatically cancel its ontological commitment when "fictional" adjectivally modifies what there is), or (2) noncommitting usages of "there is," where the ontic-canceling presence of "fictional" or its cognates is present, are easily recognized in the vernacular and distinguished from committing uses of "there is." I leave aside possibility (1) for the reasons given in my response to the metaphor/pretend option, and only consider (2).

The problem with this approach is one that we've seen before (when discussing the paraphrase option), that the real issue, the ability to engage in anaphora, isn't adequately handled by an approach turning on the cancellation of commitment by virtue of the word "fictional." Let's suppose that this would work—that it can be agreed that "fictional" always cancels ontological commitment and does so by virtue of its meaning.[53] Still, the move only faults the *example* (M) and, as a result, overlooks the large number of other uses of "there is" that speakers *clearly* don't want to be ontologically committing despite nothing like a canceling use of "fictional" occurring (even tacitly). Although discourse about fictional entities has been my primary illustration of a noncommitting use of "there are" in the vernacular because of the salient fact that, for most people, fictional objects exist in no sense at all, there are *many* other examples I could have used. Consider the following very natural inference: John dreams about ghosts; Sam dreams about ghosts; there is at least one thing that both John and Sam dream about. Or consider these examples (adapted from Hale 1987, p. 22): "There is something this form failed to record" (viz., the martial status of the claimant) or "There is something you have not disclosed" (viz., the whereabouts of the prime minister). No ordinary speaker wants to be committed to whereabouts or martial statuses, nor do most of us want to be committed to ghosts (at least not on the mere basis of the fact that others dream about such things).

---

much in the way of real confirmational impact): Early evolutionary theory required the sun to be much older than preatomic physics seemed to allow the possibility for. As we all know, evolutionary theory was vindicated, not preatomic physics.

[52] This, recall, is the view that terms like "fictional" cancel, either pragmatically or semantically, the ontological force of "there is."

[53] But does it? Consider the following sentences: "Although unicorns are fictional creatures, this doesn't mean they don't exist. Of course they do: They just don't exist in space and time—they're abstracta." I don't agree with this view, and I've argued against it. But I don't think I can argue against it on the grounds that this sentence is contradictory by virtue of the meaning of the word "fictional." However, just this claim must be made by proponents of this version of the cancellation option.

The reader may be tempted to engage in a divide-and-conquer strategy here and borrow moves from earlier approaches:[54] Use the fictional cancellation move on my fiction examples, invoke the avoidance of opaque contexts—in particular, propositional attitude contexts—and engage in paraphrase for the rest. I've already suggested why the avoidance of opaque contexts isn't a desirable move here, especially when the substitution of co-referring* terms isn't a problem. As for paraphrase, I repeat that what's required isn't a substantial and subtle paraphrasing of the offending idioms—that's desirable when regimenting discourse (and trying to eliminate commitments as Quine understands them). But as a tool in support of the triviality thesis, what's required are paraphrases easily accessible to speakers and naturally adopted. What's more likely, if you press speakers on their usage of "there is" in the above examples, and suggest they really take such things to exist? Will the speakers provide a paraphrase or simply deny your claim? I think they'll do the latter, which suggests that "there is" as ordinarily used in these cases is *not* a usage of laziness.

Here's what I've tried to do: I've not (in every case) attempted to show that the alternatives to the ontologically neutral anaphora option or even a combination of such can be *definitively* refuted. Rather, I've tried to show that the ontologically neutral anaphora option is far more likely than the other options. That is to say—and let me stress this—the claim I've attempted to establish is that the best option among those I've considered is that "there is," as used in the vernacular, isn't ontologically committing. Some might be inclined to draw a weaker conclusion: "There is" *sometimes* isn't ontologically committing. I draw the stronger conclusion on the grounds that uses of "there is" are not ambiguous (recall the discussion of the metaphor/pretend option).

Some further remarks, however, should probably be made in support of the ontologically neutral anaphora option. Some may be concerned, in particular, that this option isn't compatible with views that take the role of (genuine) singular terms to at least have to *include* the contribution, of what's referred to, to the proposition that such singular terms occur in. As I said in the second paragraph of footnote 39, I really can't get into this now because it will involve a substantial discussion of the semantics of proper names, and our concern here is only with Quine's triviality thesis. However, let me say this much: Just as in the case of objectual quantifiers, where a separation between the semantics of objectual quantifiers and the ontological force such quantifiers carry is possible, one can adopt direct reference* semantics for proper names while stripping "what's referred to" of ontological significance. That is, one can take the role of a proper name to supply to a proposition only what it refers* to, and yet that reference can be drawn from a domain (given in the metalanguage) that

---

[54] I owe this suggestion to Mike Resnik.

contains items that exist in no sense at all. Of course, a story must be told about how the sentences in which such a proper name appears are assigned truth values (a causal story, for example—which traces names back to items the properties of which determine the truth values of such sentences—isn't available). But this isn't as hard as it looks.[55]

Recapitulation

I detect (in Quine's text) two possible arguments for taking the first-order objectual existential quantifier to carry ontological commitment rather than, say, an existence predicate doing so. The first is that the quantifier does so by virtue of its semantics alone. The second is that this quantifier is a regimentation of the ordinary language "there is," and the latter (in the vernacular) already carries ontological commitment. The first argument is dispatched fairly quickly. The second argument requires a lengthy tour of possibilities because there are methods of finessing purported examples of vernacular uses of "there is" that don't carry ontological commitments. The upshot, I conclude, is that, despite appearances, Quine has no argument—at least no argument deserving the name "trivial"—for his taking the first-order objectual existential quantifier to be the sole conveyer of ontological commitment (in regimented languages).

Concluding Remarks

The mere falsity of Quine's triviality thesis doesn't refute his criterion; what's been shown is only that his criterion must be compared with alternative criteria, for example, that what a regimented discourse commits us to should be evaluated in terms of a special existence predicate minted precisely to *carry* commitment. (This last suggestion *is* the one that I'll eventually conclude—in chapter 4—is our best choice.)

This much I can say now: It *may* be argued against Quine's criterion that *our* discourse includes discourse about fiction, and that the foregoing

---

[55] It's not as hard as it looks because, in the case of nonreferring names, co-reference* between names in the object language and names in the metalanguage can be *stipulated*, and the truth or falsity of sentences involved can be a matter of the truth values we *take* them to have. This means, in particular, that names function (semantically and grammatically) the same way regardless of whether or not there's an object that they refer to. That proper names function in no semantically special way in pure mathematics is one of the major burdens of my 1994, where the view presupposed is that the semantics of mathematical language is the same as that for the rest of our language, and yet, our understanding of mathematical practice is distorted if we try to supply any content to the idea that some sort of mathematical *object* is anywhere involved in that practice. That the same must be true of discourse about fictional objects I simply take to be an indisputable datum. So as far as the semantics of proper names is concerned, the foregoing is compatible with the claim that names contribute their referents* to the propositions they appear in (and where their referents* are described in the metalanguage)—and this is so whether "what they refer to" exists in any sense at all. Recall the slogan about semantics and ontology given earlier: It applies to the semantics of *every* locution, not just to the semantics of the quantifiers.

has shown that any discourse concerned (in part) with a fictional subject matter can't be regimented by a first-order language where objectual quantifiers bear ontological commitment. This is because such regimentation requires options that have been shown to be inadequate.

It may also be argued that the indispensability of mathematical doctrine to science shows that introducing an existence predicate into regimented discourse to indicate ontological commitment is better than having objectual quantifiers carry commitment, because otherwise we find ourselves committed to the existence of mathematical objects, and we can exclude these undesirables by crafting an existence predicate that mathematical abstracta don't fall under.

These last two suggestions, however, which pit Quine's criterion against alternatives, raise new considerations, because both moves turn on an antecedent desire to avoid commitments, respectively, to fictional objects and to abstracta. It seems, that is, that adjudication among criteria for what a discourse is committed to requires antecedent *criteria for what exists*. This requires an extended discussion of its own, to which I now turn.

4

Criteria for What Exists

In chapter 3, I considered candidate criteria for what a discourse commits us to. In so doing, I followed Quine's lead, because this is how he frames the issue in his 1948 essay. Settling on a criterion for what a(n arbitrary first-order) discourse is committed to seemed, in his hands, a logically prior consideration cleanly decidable independently of tangled metaphysical arguments about what exists. But Quine's argument for his criterion turns on a sleight of hand, and once competing criteria for what a discourse is committed to are back in the race, we seem forced to worry about what criteria, if any, there are for what exists.

One might have thought that the Quinean approach (of taking a criterion—for what a discourse is committed to—to be logically prior to anything like a criterion for what exists) is misguided to begin with because the factors that mix together in theory construction to force what's quantified over in our best theories aren't *all* metaphysical ones about what's in the world; other factors, recognizably purely linguistic ones—involving notational tractability for example—play a role too.[1] Quine would agree but would claim, and in this *many* philosophers have followed him, that hopes of distinguishing these factors, except in the most obvious cases, is quixotic. A specific corollary of this more general stance

---

[1] All parties to the debate recognize this, and the standard example of a case where linguistic convenience leads to a commitment to numbers is "The number of *A*s is the same as the number of *B*s," where, say, the number of *A*s and *B*s is indeterminately large. See chapter 3, n. 50.

is the Quine-Putnam indispensability thesis that we've been so exercised over in chapters 1 and 2.

As should be clear, I disagree with this general stance, at least as far as ontological commitment (not truth) is concerned, and much of this book, especially part II, is dedicated to showing how to separate the linguistic factors in our theorizing from the impact of the world on it. In any case, the current motive in moving to criteria for what exists is more modest—the Quinean attempt to *force* the quantifiers to carry ontological commitment, either on technical grounds or because of practices in the vernacular, has broken down.

Therefore, here's the goal of chapter 4: I'll first show that the philosophical indeterminacy that we found when searching for a criterion for what a discourse is committed to, is *not* alleviated when we turn to criteria for what exists. Instead, we find that there are competing criteria here, as well, which can't be rationally adjudicated among. So I turn to the question of which, if any, criterion we've *collectively adopted*—and find that ontological independence, in the appropriate sense, is such a criterion.

This criterion for what exists, ontological independence, *could* be compatible with the Quinean criterion for what a discourse is committed to—but *isn't* given (1) that mathematical abstracta are *not* ontologically independent in the appropriate sense (something that I also establish in this chapter), and (2) that *mathematical doctrine* coupled with its *quantifier commitments* proves *indispensable* to empirical science (a claim illustrated in part II). Together, these two claims force us to break with the Quinean criterion for what a discourse is committed to, and to adopt an existence predicate instead, where such a predicate applies (truth-successfully) to those terms that refer* to what's ontologically independent or, because it's an *existence* predicate, to those (and only those) terms that *refer*.

Let's begin. There seem to be a number of candidate criteria for what exists, purported conditions, that is, that *anything* that exists must satisfy. For example, what exists must be (1) observable, (2) causally efficacious, (3) in space and time, and (4) not ontologically dependent (in certain ways) on linguistic or psychological processes. In the past, I've tentatively suggested that some of these metaphysical criteria are subscribed to (implicitly) by certain philosophers, and I've even named names.[2] This was rash, however, as some examples will show.

1. A philosopher may want to argue that the only objects we should ever be committed to are *observational* ones. And the argument may be that, in fact, it isn't that, metaphysically speaking, anything that exists *must be* observable (how on earth would one establish *that*?); rather, the claim is an epistemic one: Any *thing* we can be in a position to know something about must be something we can observe. Thus, what's of-

---

[2] See my 1998, p. 2.

fered isn't a criterion for what exists; rather, an epistemic claim is being made about what sorts of things we can have knowledge of, or about what sorts of things we're *justified* in claiming knowledge of.[3]

2. Second example: A philosopher may want to argue that the only objects we should ever be committed to are ones that in some sense we've causal access to. And the argument may be that, in fact, it isn't that, metaphysically speaking, anything that exists must have causal powers (of some sort)—how on earth would we ever show *that?*—rather, the claim is an epistemic one: Any *thing* we're in a position to know something about must be something we can causally interact with. Again, what's being offered isn't a criterion for what exists; rather, it's an epistemic claim about what sorts of things *we* can take ourselves as justified in claiming knowledge of.[4]

3. Third example. A philosopher may want to argue that the only objects we should ever be committed to are concreta or items in space and time. And the argument may be that, in fact, it isn't that, metaphysically speaking, anything that exists *is concrete*, that abstracta are, metaphysically speaking, the sorts of things that don't exist; rather, the claim is an epistemic one: Any *thing* we can maneuver ourselves into a position to know something about must be something that we've managed to causally interact with, or something to which our knowledge claims are reliably connected. And the only candidates for which *that* is possible are concreta or items in space and time.[5]

Contrast these sorts of arguments with the one about fiction that I closed chapter 3 with: It was suggested that fictional objects don't exist, not because such objects are ones that it's hard to find things out about; rather, it's all too easy to find things out about fictional objects because authors simply *make the facts up*. It seems, therefore, that the denial of the existence of fictional objects turns on a metaphysical claim, something to the effect of, *What's totally made up doesn't exist.*[6] That is, a necessary condition has been placed on anything that exists: it can't be totally made up.

So examples (1) – (3), as described, aren't, strictly speaking, staking out *metaphysical* positions; rather, they're claims that (because of how beings like us must gather knowledge) we can't know *truths* about certain sorts

---

[3] Some such view as roughly described *may* be attributed to van Fraassen (1980).

[4] Some such position as roughly described *may* be attributed to Cartwright (1983), and *perhaps* also to Hacking.

[5] Some such roughly described position *may* be attributed to certain nominalistically inclined philosophers of mathematics, e.g., Field (1980).

[6] Fictional objects aren't like artifacts, sculptures, for example. There is something—call it stuff—that sculptures are made *out of*, and this is true of artifacts in general. *Fictional objects* aren't made out of anything. (It's confused to think fictional objects are made up of, say, words. It's similarly confused—but perhaps a little harder to see why—to think that movie characters are made out of, say, celluloid, or that painted characters are made out of paint. Of course, what a sculpture *depicts* needn't exist any more than fictional objects do.)

of purported objects. And so, on one construal of examples (1) – (3), the central targets are *truths* about such objects: We can't take ourselves to think that certain sentences (about nonobservables, about causally inaccessible items, about abstracta) are *true*.[7] As I showed in chapters 1 and 2, however, this position is untenable, because regardless of *how* certain statements are established, unless they're shown to be dispensable to scientific practice, or to have implications that are quarantined, we're committed to their truth (and no matter how much this grates on the epistemic sensitivities of some philosophers).

We can, however, reconstrue examples (1) – (3) as making claims *not* about what sentences should be taken as true or not on the basis of our epistemic practices but, rather, as making claims about what entities we should take ourselves to be committed to. Examples (1) – (3) may be described, that is, as trying to mark out the *instrumental character* of certain sentences in our conceptual scheme, not by identifying them as false or as truth-valueless or as statements the truth-values we're ignorant of, but as containing terms that refer to objects that don't exist (refer to nothing at all).

Here's how to so reconstrue example (1): Use a predicate, "observable" ("$\mathbf{O}_b$"), and then argue that a sentence $(\exists x)\mathbf{S}x$ is *purely instrumental* (even if true) if $(x)(\mathbf{S}x \rightarrow \neg \mathbf{O}_b x)$ holds. Argue further that we commit ourselves to a class of items $\mathbf{S}$ only if the sentence $(\exists x)(\mathbf{S}x)$ & $(x)(\mathbf{S}x \rightarrow \mathbf{O}_b x)$ is an implication of our discourse.[8]

It may seem that substantial debates in epistemology are in the offing. For example, one may argue that scientists do take observation of something to be the only appropriate grounds for regarding that something as (provisionally) existing: They're not willing to be committed to a kind of entity unless they can observe it (in some sense), and thus, even though scientists take, for example, mathematical claims or claims about theoreti-

---

[7] This is how I understand the positions, respectively, of van Fraassen and of Cartwright and Hacking. Field's program is similarly motivated: Show that applied mathematical sentences are dispensable instrumental tools, and so, that mathematical predicates can be taken to hold of nothing, that mathematical constants can be taken to refer to nothing (and thus, mathematical sentences can be assigned truth values according to this null interpretation of their predicates and constants).

[8] I.e., only in this way are we committed to *every* $\mathbf{S}$. $(\exists x)(\mathbf{S}x$ & $\mathbf{O}_b x)$ merely commits us to *some* $\mathbf{S}$s existing.

There seem to be two possible construals of the ontological interpretation of $\mathbf{O}_b$: On the first, it's not acting—strictly speaking—as an "existence" predicate but rather as an "epistemically licensed to regard as existing" predicate; on the second, it *is* an "existence" predicate, but (of course) like every predicate, (1) we can be wrong about what that predicate (actually) applies to, and further (in the case of predicates we invest with ontological significance), (2) we can be wrong about *which* predicate we should treat as ontologically committing. (Perhaps some of what we observe doesn't exist, or perhaps many things we don't observe do exist.) For a philosopher who thinks that, anyway, we can be wrong—pretty much—about *anything*, these two construals seem to come to the same thing. Despite this, I give reasons at the end of this chapter for why the "epistemically licensed to regard as existing" option for any candidate existence predicate is *not* a good choice.

cal entities to be *true*, they're not committed to the existence of the things such statements seem to be about. Or one may claim—in opposition to the foregoing—that the epistemology of ordinary science doesn't operate the way the observationalist thinks it does: Instrumental access of the right sort to theoretical entities is routinely seen by scientists to be every bit as good, epistemically, as observation itself.[9] The proponent of the latter position may join forces with the observationalist, however, when it comes to mathematical abstracta: We don't see *those*, and we don't instrumentally interact with them either, so no epistemic license to regard them as existing is available on either view.

Despite this anticipation of philosophical struggle in epistemology, we're unlikely to get anywhere (philosophically speaking) if the only considerations raised are epistemic ones of this sort, and this is because the epistemically reconstrued positions lack the apparent force of the metaphysical originals. For if statements about nonobservables, about causally inaccessible items, about abstracta, must all be taken to be *true*, what prevents someone from arguing that (clearly) these items are ones that *can* be learned about the way we (in fact) do learn about them? Nonobservables, causally inaccessible items, abstracta, are all what we might describe as "epistemically lightweight": They're items we can learn about *without* use of (respectively) observation, causality, and so on. (And the fact that we must take certain statements about them—and not others—as *true* shows this.)

One might try to blunt the force of this last claim by invoking Occam's Razor, by arguing that if it isn't necessary for us to commit ourselves to such epistemically lightweight objects, then we shouldn't. Notice that the needed form Occam's Razor takes here is as a principle that adjudicates among rival descriptions of what entities we're epistemically justified as taking to exist on the basis of a commitment to the *truth* of a discourse.

Contrast this with the form that Occam's Razor takes when a criterion for what a discourse commits us to is already accepted, Quine's criterion, for example. Then we compare two theories otherwise alike (in simplicity and in empirical adequacy, say) and choose the one that commits us to the fewest types of entities. Here, instead, we're evaluating two ontological characterizations of a single discourse, one that takes that discourse to be committed to abstracta, and so on, despite their epistemic lightweightedness, and one that doesn't. Unfortunately, any such attempt to invoke Occam's Razor in this form is premature, because there are no epistemic principles (or metaphysical ones, for that matter) yet in place to base it on. If, for example, we had epistemic considerations already in hand that revealed *which* items referred to in a discourse are ones we're (epistemically) justified in taking to exist (on the basis of the evidence that established the truth of the statements in that discourse), we would, of course, be able to evaluate two descriptions of the purported ontologi-

---

[9] See my 1997b.

cal commitments of a theory by comparing them epistemically, and then we could invoke this form of Occam's Razor against that description that took that discourse to be committed to more than was epistemically justifiable. Thus, if we already knew that only causal accessibility to a type of entity sufficed for knowledge of any sort of entity, we could rule out a description of a discourse's commitments to a type of entity on the grounds that the type of entity was causally inaccessible. Thus, it seems that the required form of Occam's Razor isn't something we can use to justify epistemic conditions on what we're justified in taking ourselves to know exists, but something that, once such conditions are in place, can be invoked on their basis.

Perhaps, however, this is too quick. The careful way to proceed here (one may think) is to examine how the truth of sentences gets established. Having done so, one can see that certain entities play a *role* in the establishment of the truth of the sentences, and that although other items may be *referred to* in the sentences established as true, in the technical sense that there are noun phrases in the sentences that purport to refer to such entities, we aren't justified in taking these other items to exist because they *don't* play a role in the evidential process establishing the sentences. Put it this way: We can ask what entities *need to exist* to explain how we come to know what we know. And only those entities so needed are the ones that we're epistemically justified in taking to exist on the basis of how the truth of a set of sentences was established.

Consider this sentence:

(A) There are the same number of toads in my garden as there are cats.

The argument is that to establish the truth of this sentence, we need to count toads and cats, but in doing that, we don't interact with numbers. There are different ways of cashing out this notion of interaction, but two that come to mind (to my mind, anyway) are (1) a causal notion—we causally interact with toads and cats but not with numbers; or (2) an observational notion—we see toads and cats but not numbers.[10] Both characterizations of the sort of access to entities that's essential to our knowing truths can be generalized nicely to theoretical entities such as subatomic particles or black holes,[11] but no respectable extension of these ideas reaches abstracta such as numbers. In explaining how we came to recognize the truth or falsity of the claim that the numbers of toads and cats in my garden are the same, access to actual toads and cats is required in that explanation; the same isn't true of numbers.

---

[10] The second claim about how to distinguish epistemically significant differences between our transactions with abstracta and our transactions with concreta is open to debate. See chapter 5. Because the move fails in any case, I'm granting (temporarily) that the distinction can be made out in one or another of these ways.

[11] Causal access to entities nicely generalizes from sensory access to objects to include instrumental manipulation of items out of our sensory range. So, too, observation generalizes to what I call "thick epistemic access." See my 1997a, 1997b, and chapter 6.

It's important not to beg questions. What is it about the fact that toads are interacted with (and numbers aren't) that makes the number 7 (say) inessential to the epistemic process that established (A), and toads essential to it? The thought is this: If, for all we know, there are no numbers, it doesn't change anything about the events in the garden that established the truth of sentence (A).

This last point, it must be said, amounts only to the truism that numbers neither participate in causal processes nor are the sorts of things that one can observe (this is why the possible nonexistence of numbers has no impact on those processes). But it's hard to see how any of this shows what's needed. For why can't this entire line of thought be challenged like so: *Why* is it a requirement, on the sorts of objects we can be justified in knowing exist on the basis of the truth of a discourse, that only those items that participate *in the evidential process itself* (either by being observed or by causal involvement) are candidates? If numbers exist, and have properties knowable by the methods we use to establish truths about numbers, why isn't it just irrelevant that they don't participate in any way in the epistemic processes that established those truths?

Well, one might say plaintively, we're *forced* to believe in cats and dogs, and we're not forced to believe in numbers by the epistemic process of establishing the truth of (A). But again, so what? And anyway, why can't it be said that we're forced to believe in numbers by our belief in the *truth* of sentence (A)?[12]

There seems no way around it. We're forced to an excursion into *metaphysics*—specifically, into a study of necessary and sufficient conditions for what exists—because only then can we raise (and perhaps answer) the question of whether *things that exist* can be the sorts of things that we can learn about in these various (epistemically lightweight) ways. Let's, therefore, consider various candidate criteria (necessary *and* sufficient conditions) for what exists and address the question of how to adjudicate among *them*.

Before doing this, we must guard ourselves against potential fallacies due to the trickiness of reasoning about nothing at all. Consider, for a moment, the nature of necessary and sufficient conditions. Given, as I've argued, that things that don't exist don't have *any* properties, it seems to follow pretty cleanly that if a condition A is necessary for the existence of something (if, i.e., anything that exists must have the property A), it

---

[12] There are echoes here of an earlier debate in epistemology, for the above is just another example of the common temptation to restrict knowledge claims to the purported total evidence *for* such claims. So, e.g., if we think that *all* our evidence for anything consists of observations, we can be tempted to think that we're only justified in knowledge claims about the existence of observable entities.

I should add that during the course of this chapter I'll eventually settle on the position that numbers exist in no sense at all, a position that someone plying this verificationist-flavored argument will no doubt find appealing. But, as the above indicates, I can't accept a verificationist route to that position and must employ very different arguments.

straightaway follows that A is both necessary *and* sufficient for existence. For sufficiency can be ruled out only by it being possible for something that doesn't exist to have the property A, and, as just stated, things that don't exist have no properties. It may also seem that talk of a property being *sufficient* for what exists is, in any case, hard to make sense of in a way that goes beyond utter triviality. For a condition A is sufficient for the existence of something, if given it holds of that something, then that something exists. But if condition A does hold of something, then, of course, that something has a property (A, in fact) and so is already presumed to exist.

The above piece of reasoning strikes me as fine; the issue it raises also strikes me as analogous to one that arises about identity. Consider the following famous consideration: $a = b$ can't be genuinely informative because it can't be informative that something is the same as itself.[13] However, what makes the term "=" so *indispensable* is *not* that it picks out a metaphysically significant identity that we periodically need to reassure ourselves (aloud) about; rather, what makes it important is that nothing in a pair of names indicates whether they co-refer or not: Their co-reference must be stated explicitly. This fact about usage makes it tempting to think that "$a = b$" is a metalinguistic statement about the *names* "a" and "b," rather than, as it appears to be, a statement about the object(s) those names refer to. The fallacy in this last bit of reasoning, however, is the (implicit) assumption that if a statement is useful because it sorts out facts about our usage of *terms*, it therefore must be *about* those terms.

Something very similar happens with a genuine existence predicate if we allow, which I do, the statement of truths "about" fictional entities and abstracta, even though such things are simultaneously claimed to exist in no sense at all (and thus not to have properties *either*). To say that Mickey Mouse is a fictional mouse, for example, isn't (really) to attribute a *property* to *Mickey Mouse* because, after all, things that don't exist have no properties.

Here, too, there is nothing to object to. The semantics of our language, even if that semantics is understood as involving first-order objectual quantifiers, doesn't require propertied nonexistent objects in order to make true statements with noun phrases that don't refer. It's tempting, no doubt, to notationally distinguish attribution of properties to things that exist from the "fake" attribution of properties to things that don't exist—but logically and metaphysically speaking, there is no real reason to do so. The presence of an existence predicate alone suffices to distinguish those cases where the attribution of a property is to something real and where a sentence of exactly the same syntactic and logical form must be understood (metaphysically) in a different way: as a sentence that is

---

[13] Wittgenstein's (1922) dictum, 5.5301, on this is even more harsh: "Roughly speaking, to say of *two* things that they are identical is nonsense, and to say of *one* thing that it is identical with itself is to say nothing at all."

true, but not because it describes (correctly) that something has a certain property.

However, when speaking, metaphysically, of *criteria* (necessary and sufficient conditions) for what exists, a (benign) split arises between how these conditions apply to what exists and how these conditions arise as applied to our sentences, and this split is similar to what happens with identity. As far as (real) objects are concerned, what there is is only what exists, and because there isn't anything else, and so, nothing else that has properties, talk of necessary and sufficient conditions on what exists seems strangely trivial (any property is sufficient for existence if it's instantiated, otherwise not; any property that is necessary for what exists is sufficient for what exists).

But given a language with nondenoting noun phrases—noun phrases that pick out nothing at all—and given that predicates can apply to such terms just as they apply to any term, and yield truths, necessary and sufficient conditions may exist for the existence predicate, that is, conditions of the sort $(x)(\mathbf{A}x \Leftrightarrow \mathbf{E}x)$, where "A" picks out some condition or other ("causally efficacious," "located in space and time," etc.) and "E" is the existence predicate. The language allows the cogency of such a condition "A" being co-extensive*[14] with the existence predicate, even though, strictly speaking, it's only terms that denote nothing that the co-extensionality* of "A" with "E" excludes from yielding truths when appended to either "E" or "A," and not the exclusion of the application of "A" to ghostly nonexistent objects; further, this can be a *significant* claim to make because "A" may be distinguished from other predicates "B" and "C," which apply to nondenoting noun phrases to yield truths, and thus can nontrivially fail to be necessary for existence, sufficient for existence, or both.

This take on the cogency of necessary and sufficient conditions for existence raises, however, what I'll call *Kant's challenge*:[15] that *any* list of properties can apply to something that doesn't exist, and so *no* property can be co-extensive* with "exists," when construed ontologically.

I don't understand "Kant's challenge" as a *metaphysical claim* about what doesn't exist, and how properties can be attributed to such things; rather, it's a claim about how the predicates in our language operate, so that it amounts to the claim that, in fact, there is no predicate or set of predicates **A** for which the assertion $(x)(\mathbf{A}x \Leftrightarrow \mathbf{E}x)$ can be made. The reason for this, if it's true, lies in how we allow ourselves to assert truths

---

[14] Recall the notions of co-extensionality* and reference* from chapter 3. What's pertinent to the relationship between a predicate **A** and the existence predicate **E**, if the former is to function as a criterion for the latter, is of course not just that they're both co-extensional (on (all) the same (real) items), but that they *both* fail to yield truths when applied to all nondenoting noun phrases, i.e., that they're also co-extensional*.

[15] See chap. III, § 4 of Kant's (1787) Transcendental Dialectic, pp. 500–7, esp. the cute bit on p. 505 about the thalers. Of course I'm *not* claiming that Kant's challenge, as I construe it, is what Kant had in mind. Also see Hume 1739, bk. I, pt. II, § VI.

involving nondenoting noun phrases: The claim is that for any predicate "**A**," which is a candidate for co-extension* with "**E**," there is some nondenoting noun phrase *n*, where **A***n* is true.

This claim may look plausible: Consider *any* predicate you like. Surely a piece of fiction is possible in which that predicate is attributed to *something*. Consider the list of potential criteria for what exists that I offered at the beginning of this chapter: "Observable," "causally efficacious," "in space and time," all surely apply to Sherlock Holmes. He lived in London, didn't he; and he could be seen on many days by Dr. Watson, isn't that true, too? Even "not ontologically dependent on linguistic or psychological processes in certain ways" seems true of him—doesn't it?—because he wasn't a hallucination of the other characters, something not true of *everyone* in fiction.[16]

But this last suggestion gives the game away. I chose my examples of truths about fictional objects in chapter 3 carefully. I'm willing to accept as true the claim that "Sherlock Holmes is admired by many real detectives"; I'm not willing to accept as true the claim that "Sherlock Holmes lived in London in the late nineteenth century." We must distinguish *true* claims about fictional objects from those only pretended true. What takes place *in* fiction (usually) is pretended true, no doubt about that. This doesn't force *every* claim about fictional objects to be only pretended true (as I argued in chapter 3), but this is enough to meet Kant's challenge. It's simply not true that fictional beings or other denizens of the nonexistent, such as hallucination beings, dream beings, and so on, are causally efficacious or not ontologically independent of us in the appropriate respects.[17]

That abstracta, the other class of items that I'll argue exist in no sense at all, palpably have predicates holding of them that do *not* hold of concreta is, I dare say, obvious.[18]

---

[16] It's not true, e.g., of Cale in Jonathan Lethem's *Amnesia Moon* (1995).

[17] Are dream beings *observable*? Are hallucination beings in space and time *and* observable? I'd hate to have to argue that these predicates *don't* apply truthfully to noun phrases referring* to such things, given how we talk. Luckily, it doesn't matter to me one way or the other. For Kant's challenge to succeed, it has to be that *no* predicate exists that can be used to distinguish what's real from what's not. To deflect Kant's challenge, therefore, I need only find at least *one* predicate (other than "*really* exists") that yield truths when applied (to all) things that exist but does not yield *any* truths when applied to nonreferring noun phrases: *My* candidates are "causally efficacious" and "ontologically independent"—the latter in the appropriate sense.

[18] Being a claim about our language, it's possible that Kant's challenge could have succeeded—after all, couldn't our language have evolved so that any predicate applicable to terms that refer could also have been applied to nonreferring terms (of one sort or another) and yield truths? Well, yes, of course that's in principle possible. I offer some speculations later in this chapter for why things didn't go this way.

I should add that the reader shouldn't overestimate the significance of my needing a response to Kant's challenge: It wouldn't have changed very much if that challenge *had*

Two last points before returning to the main theme of this chapter, criteria for what exists: First, the temptation (stylistically speaking) is overwhelming to descend semantically when speaking of the nonexistent, to say, for example, that "it's not the case that, given any property, there is some item that exists in no sense at all which has that property," rather than "it's not the case that, given any predicate, there is some nondenoting noun phrase that when appended to that predicate yields a truth." I hope a cursory comparison of the two sentences just quoted explains *why*. The related considerations raised in chapter 3 explain why the temptation to speak directly of nonexistent objects and their properties rather than go systematically metalinguistic is so natural; I hope I have also made the case there that doing so does *not* (at least on these grounds *alone*) involve ontological commitments of *any* sort. Lastly, given the delicate nature of the discussion just engaged in, I've tried *very hard* to couch it almost entirely in metalinguistic terms to avoid misunderstanding. I hope (hope is a major theme of this paragraph) that the reader forgives me for not being rigorously consistent in this regard. I'll continue to fail to be rigorously consistent in this regard: I'll sometimes speak of nonexistent objects and their properties in what follows. No more than before should such talk be seen as ontologically committing to things (of *any* sort) with properties.

Second, although in discussing criteria for what exists we're implicitly engaged (much of the time) in a discussion of the properties of *our* language, I'll nevertheless continue to describe this as an analysis of *metaphysical considerations* to contrast it with the previous *epistemological* ones.

OK. Let's return now to the question of what sorts of candidates we actually have for necessary and sufficient conditions on existence. If we go back to that list of candidate criteria I opened this chapter with, it doesn't seem that any of them are items that one can argue successfully for *on purely metaphysical grounds*.[19] Take observationality, for example: As anyone with eyes can tell you, observationality is a deeply parochial feature of the observing creature. It's hard to believe that anything that

---

been sustained. After all, whoever said that a philosopher (me, for instance) couldn't *coin* terms "causally efficacious*," "in* space and time," etc., stipulated to apply truth-successfully *only to* terms that refer?

[19] There are two general reasons why attempts to provide genuine metaphysical arguments for constraints on what exists prove elusive. Briefly, "exists," like "true," "gold," and other terms, is criterion-transcendent. It's a word that's averse to fixed criteria that constrain what terms it can apply to in order to yield truths. See my 2000b, pt. 4, § 3. The second is even more significant: It's that we can't—on purely linguistic grounds—claim that the word "exists" itself *always* indicates ontological commitment. This should already be clear from the fact that the use of what should be merely a rhetorical enhancer, "*really* exists," does *not* sound redundant to the untutored ear. See chapter 5 for further discussion on this.

exists *must* be something that *could be* observed; it's hard, anyway, to see how something like that could be established without a substantial analysis of observation itself. Certainly, it looks like an entirely empirical question whether what could be observed is co-extensional* with what exists; it *doesn't* look like something that can be established by merely treating observationality as a *criterion* for what exists.[20]

Causal efficacy isn't much better. *Why* should anything that exists be causally efficacious? Why couldn't (some) things be causally inert? Notice that to argue that we can explain what needs to be explained about the universe on the basis of what's causally efficacious, and that therefore by, say, Occam's Razor, only such items exist, isn't to provide an argument of the desired form. The latter is an *epistemic* argument based on the idea that we need only commit ourselves to what we need (ontologically speaking) for explanation. But what relevance has this epistemic point to a criterion for what exists? How does it show that what exists must be causally efficacious (vs. showing the weaker claim that on epistemic grounds we need never find ourselves committed to anything that isn't causally efficacious to explain what we need to explain—an argument, as I've already shown, that won't work on its own)?

Location in space and time is no better. (How do we establish that what exists *must be* in space and time? Why doesn't this, in the crassest way imaginable, simply beg the question against the traditional Platonist?)

The fourth candidate criterion, that what exists must not be ontologically dependent on any linguistic or psychological process, is also troubled. However, it's a candidate criterion for what exists that, on the face of it, is more plausible than the others just considered. Furthermore, it's a candidate that, under certain descriptions, for example, "mind independence," philosophers have been *metaphysically* motivated to adopt as a criterion for what exists (vs. having a motivation really due to epistemic considerations).[21] The reasons, therefore, for why there is no (purely metaphysical) argument for its being a constraint on what exists deserves a more extended discussion.

It's not as easy to characterize what's meant by claiming that something is "mind dependent" as it is to give examples.[22] The most straight-

---

[20] Also see n. 17 with regard to this and with regard to the criterion: location in space and time.

[21] See Vinueza (2001), and the literature he cites. Also see my 2000c.

[22] The problem with producing a crisp definition is that this notion seems to include too much. Minds, mental states, language, speech acts, works of art, and artifacts are all items we're prone to think *do* exist but that also seem to be mind dependent or ontologically dependent on us in some sense. See Vinueza 2001.

One suggestion Vinueza doesn't explore, and that seems not to have been previously considered in the literature, is to presuppose *something like* a causal theory of reference as already in place. We can consider a term to, at best, pick out something ontologically dependent on us in the desired sense if it fails to denote anything *causally speaking*. Ontological dependence (in *this* specific sense) then emerges as an extremely natural necessary

forward ones are these: characters in dreams, imaginary beings that one thinks about, fictional characters, myths, and so on. And among *these*, the most straightforward examples are fictional characters designed from scratch by authors for the first time. The idea is that in deciding what's true about fictional entities, the author isn't required to square what's attributed to such entities with anything: He or she can "make it all up." Perhaps it can even be said that what's true about such fictional entities is only what the author *stipulates* as true of them. And something all of whose properties are ones stipulated as true isn't something that is "mind independent" or "ontologically independent of us."[23]

Suppose something is dependent on our linguistic practices in exactly the way that fictional entities have just been described as being. Why *must* it be that such objects don't exist? What sort of argument could be offered for this metaphysical claim? That is, why *couldn't* fictional practices be ones that commit its practitioners to entities—entities, admittedly, that exist in their own fictional worlds, but that nevertheless exist? Is there anything philosophically conclusive we can say about *sheer* existence that shows it's inconsistent with the "mind dependence" or the "ontological dependence" of fictional entities? No, there isn't, as the following (extended) example makes clear.

Imagine two communities: The first (us) takes fictional beings to exist in no sense at all (we're *naive fictionalists*); the second community practices fiction exactly the same way we do but does take fictional beings to exist—in as robust a sense as anything else exists—although they think fictional entities (of course) exist in their own (fictional) worlds (they're *fictional realists*).

---

condition for what exists because we can say, as we'd like to, that such terms fail to denote anything *at all*. We can then beef it up to a necessary and sufficient condition by generalizing to any possible language whatsoever (not just ours). Although I find this suggestion appealing, an exploration of whether it can be ultimately made to work raises issues that can't be pursued now.

[23] I should point out that even in this case the author may be constrained in what he or she is allowed to imagine—by concerns with what the public will like, the author's own taste, etc. But no constraints arise from the properties of the object being described. As I said, this is the easiest case. Dream beings, hallucinations, or cases where authors write about characters from other books or from mythology, etc., offer something more by way of complication. In particular, the use of the term "stipulation" is less apt when involuntary psychological events such as the contents of hallucinations or dreams are involved, or when the fictional entity is one whose characteristics are publicly known. However, the tidying up that's needed to explain the desired sense in which all these items are ontologically dependent on our linguistic practices, although a little more complicated, isn't difficult.

Some of these items can be described as "language-created" in Schiffer's (1996) sense. He takes himself to be sketching out a middle position according to which such "language-created" items are ontologically deflated because they've no "hidden and substantial natures for empirical investigation to discover" (p. 159), beyond what our language practices establish them to have, but nonetheless are items that exist unaccompanied by language-creating entities in *other* possible worlds. Later in this chapter and in chapter 5, I give reasons to reject this compromise ontological position.

At least two views of how fictional realists view fictional existence are possible.[24] The first requires "fictional space," a metaphysical haven of already existing fictional worlds in which every fictional possibility eternally resides. When someone writes a novel, say, every term in that novel refers to the beings in at least one fictional world that matches perfectly the descriptions in that book. Indeed, the terms in the novel actually refer to the events and beings in more than one fictional world—they refer to fictional objects in infinitely many such fictional worlds—because a novel leaves incomplete many details that are completed in various ways in the various fictional worlds.[25]

This first view faces serious problems. Consider an alternative: There *are* such fictional worlds, but fictional terms refer to nothing in any of these because fictional terms don't refer at all. Or (a worse possibility), fictional terms refer to fictional beings in such a way that nearly everything novelists say about their characters, and what happens to them, is *wrong*. On what grounds can we adjudicate between these views? There is nothing in the second community's practices that fixes what its fictional terms refer to in such a way that evidence can be brought to bear for and against these alternatives to decide among them.

What if the fictional realists in the second community *claim* that their fictional terms pick out fictional worlds as described? Isn't *that* evidence? So what if they *claim* this? Why couldn't they be *wrong*? Notice that not only is there nothing about their fictional practices that can be brought to bear successfully on the question of what they refer to, but it's not even imaginable what facts could *ever* arise to adjudicate among these alternatives. (That no data, even in principle, bear on this question one way or the other suggests not only that the fictional beings are made up, but also that the referential story has to be made up, too. But why think a "made up" referential story bears any resemblance to genuine *reference*?)

Some may argue that because this view relies on a "description theory" of the reference of fictional terms, it doesn't allow "referential mishaps" (terms pick out *exactly* what the descriptions associated with them pick out), and so the concern about these alternative possibilities is an unreal one. But this response overlooks a worse problem, which is that this approach can't guarantee, of the fictional objects taken to be independent of us, that they have the properties needed by the descriptions

---

[24] Views that separate what's attributed to fictional objects (while practicing fiction) from the properties such objects have (see the discussion of option (2) in chapter 3) elude the specific objection to the "fictional space" view that I raise, but face easily constructed generalizations of that objection. In any case, my aim here isn't to evaluate the cogency of every realist position about fictional nomenclature, but to find *one* that's immune to philosophical attack on metaphysical grounds.

[25] Alternatively, one might claim that fictional worlds (and fictional objects) are "incomplete objects" individuated precisely by the predicates attributed to them, and no others. In this case, the terms in a novel refer to the items in one and only one fictional world. This latter view is similar to that taken of nonexistent objects by Parsons (1980).

we forge to pick them out. This problem persists even if, on this view, *all* fictional possibilities (in some logically broad sense) are present in fictional space. For that space, and all the possible fictional worlds within it, are still taken to be ontologically independent of us, and this means we've no guarantees that the wealth of possibilities that we need (given our capacity to generate fictional descriptions) are *out there*.[26]

The second view is that fictional objects are brought spontaneously into existence by their authors in the process of creating fiction; not only that, but the traits of such objects change as authors change their views about them. (Revising a novel has metaphysical consequences!) In this case, the objects are regarded as stipulated in the sense that the properties they *actually have* are ones they're stipulated to have by their inventors. This view of fictional realism doesn't face the referential problems that the first view faced because there are no antecedently existing entities that fictional terms are supposed to refer to. Instead, the properties of the objects are dictated by the very things said about them.[27]

Given the second view of the metaphysics of the practice of fiction by the second community, ontological debate between us and them is

---

[26] I raise the same objection against Ballaguer's "full blooded Platonism." See my 2000c, pp. 238–42. It's important to realize that this objection can be pressed against *any* view that (1) takes a collection of objects to be correctly characterized by a set of descriptions *we* take to be true of them, (2) nevertheless takes such objects to be ontologically independent of us, i.e., to be items we can't stipulate the properties of, and (3) uses *only* logical or conceptual resources of some sort to bridge the (epistemic) gap between (1) and (2). See the discussion below of nontrivial explanations of the satisfaction of the reliability requirement.

Deutsch (2000) speaks of a "fictional plenitude [that] is defined by a logical comprehension principle that says, in effect, that the domain contains any and every variety of object and event imaginable—and then some" (p. 155). Further, "a description recorded in the course of making up a story cannot fail to describe something and it cannot be a misdescription of anything" (pp. 155–6). We seem, on this view—in fiction, anyway—to have the remarkable power to stipulate both the contents of the fictional plenitude and to refer to those contents. Such things, therefore, *cannot* be ontologically independent of us, but must be "made up" by us in the strongest sense of that phrase. Deutsch seems to make it clear that he doesn't take his descriptions of the fictional plenitude to be describing items that exist in no sense at all. He describes comprehension principles as "ontological" (p. 150), and he opposes his view explicitly (p. 155) to one that takes fictional terms to refer to nothing. On the other hand, he claims that he has "little or no commitment" regarding what the things in the fictional plenitude are, suggesting that his general view is even compatible with "something of a nominalist approach" (p. 157–8). Finally, he describes himself as seriously committed to the idea that one makes things up "out of whole cloth," or out of "thin air" in fiction, although he glosses this as recording descriptions "that are to be understood (philosophically) as describing elements of the fictional domain" (p. 167). Regardless of the truth or falsity of my views about ontological independence, I can certainly still deny that Deutsch can have *everything* he wants.

[27] Thus, the previous objection against the first fictional realist view, of how the community can be sure that what it says about fictional entities is true, is easily answered: What it *says* dictates what properties those entities *have*, and so of course what's said is true about *them*. The properties of these entities, and even the identity conditions among them, are seriously incomplete, of course, depending, as they do, *only* on what's *said* about them.

fruitless, because we agree on all the facts available to adjudicate the debate: We both agree that fictional entities are "made up" by the authors that invent them ("but how does this *show* the entities don't exist?" fictional realists ask us). Nor is it possible to show that the fictional realist gets into some sort of trouble by claiming that fictional objects exist. It isn't, after all, that the fictional realist thinks this implies that an author could get *wrong* what he or she writes about fictional characters (fictional entities aren't *that* sort of entity). Nor do they think that their claim that fictional beings exist has any impact on what they say about any other kind of entity. To the objection that fictional realists are lacking a robust sense of reality, they respond that, on the contrary, they have an *exquisite* sense of reality, for they never confuse fictional entities with other sorts of entities or think that the epistemic practices used to discover things about nonfictional objects are relevant to knowledge claims about fictional objects. Indeed, fictional realists are even willing to grant that the properties fictional objects have are stipulated by their authors. But what, they ask disingenuously, does the fact of stipulation have to do with *existence*?

Both fictional realists and naive fictionalists agree: If one uses language to refer to a made-up character, then just because a real person shows up with the same properties, it doesn't mean that one was (*against one's will*) referring to that person rather than to the made-up character. Indeed, both communities have straightforward methods for preventing problems should such things happen. Suppose, for example, that *B* makes up a fictional character for a screenplay, and then someone from Ohio shows up with a lawyer because *B* has inadvertently described his life, looks, and personality entirely. Luckily, *B* (in *either* community) can protect herself against such eventualities by writing a claim of intent at the end of screenplay: *The characters in this work are fictional, and any resemblance to anyone, living or dead, is purely coincidental.*[28]

Of course, this disclaimer isn't relevant if we think that a work is fiction and learn instead that the author was actually engaged in nothing of the sort. And, of course, vice versa is possible too: We can think a character mythological and learn that he isn't. This wouldn't be a matter of our discovering that there is someone with characteristics similar to the mythological being; rather, this would be a matter of our learning that a group of people were *referring* to him. In all these cases, what has happened is that we're mistaken about the linguistic processes at work. In the first two cases, the individual is pretending to engage in one sort of activity (with one set of conventions) but in fact is engaged in another. In the third case, we're under the impression that the language describing the purported mythological being arose the way mythological language usually arises—through religious misinformation, say—whereas instead,

---

[28] Notice that this disclaimer, on the face of it, anyway, takes no position about the ontological status of fictional entities.

it's a sober description (say) of an unusual person that all the speakers actually know. Both communities have access to these ways of adjudicating among these cases; existence, one way or the other, doesn't come into it at all.

If the foregoing is correct, then the last candidate criterion for what exists, that something exists if and only if it's independent of linguistic and psychological processes, isn't any more of a genuine *criterion* for what exists than the first three are, because it fails to even be a necessary condition. Does this mean that it's truly indeterminate whether fictional entities exist or not? In one sense, the philosophical sense, the answer is "Yes." Should fictional realists stubbornly insist that fictional entities do exist, and should they invoke a version of the second fictional realist story, there is no philosophical argument—no version of Occam's Razor, for example—available to nudge them from their position. What *is* true, however, is that *our* community of speakers, for whatever reason, does *not* take fictional entities to exist in any sense at all. One can give sociological reasons for why our folk ontology took this route, but such reasons won't be philosophically binding.

Here's an example of what such an explanation might look like: The first step is to note that the word "exists" plays a *contrastive* role in local discourses—we use it to contrast actual horses with flying horses, to contrast even primes greater than 2 with odd numbers, to contrast the ether with black holes. That its contrastive role is local explains why the word "exists" doesn't have necessary and sufficient conditions attached to it (we can't predict ahead of time which contrasts we may need to make in a discourse, and how). Notice that this view bears a strong resemblance to views that take ontological positing in mathematics, for example, to have originally arisen by a metaphorical extension of the ontological positing of ordinary life. I agree for the sake of argument that an extension in usage has occurred; I disagree that there is a respectable sense in which it can *still* be labeled "metaphorical."

The second step goes like this: We're willing to extend the word "exist" to a collection of noun phrases in a discourse, *provided* the extension doesn't breed confusion in already established discourses. Talk of distinct fictional worlds is a rather sophisticated, and recent, construal of the ontology of storytelling; one not available to the general fiction-listening public several centuries ago. But without distinct worlds to segregate fictional entities within, one must distinguish them from the flora and fauna one is likely to meet in one's travels. And this motivates the commonly held intuition that they don't exist (in any sense at all).[29]

What grounds have I for claiming that *our* community of speakers doesn't take fictional entities to exist in any sense at all? The naturalness of statements like this:

---

[29] This addresses the question raised in the second paragraph of chapter 3, n. 13: The use of "exists" in mathematical discourse is contrastive in what's obviously a local discourse.

Although there are cities in H.P. Lovecraft stories that exist, there are places that don't exist that he depicts as being located in those cities.[30]

Our general acceptance that fictional objects exist in no sense at all isn't a brute intuitive fact *about fictional objects*. Rather, this intuition is an application of a more general intuition that if something is entirely "made up" or is ontologically dependent on our linguistic practices or psychological states, then it exists in no sense at all. "Ontologically dependent" here is *not* understood in the sense that it's, say, a psychological state or a linguistic item (e.g., a word)—these things *do* exist—but in the sense that it's (part of) the *content* of such a thing, and this isn't *content* in the sense that an *actual* apple is the *referent* of the word "apple," but in the more elusive sense that a hallucination of an elf (or apple) has as its content "an elf" (or "an apple") that exists in no sense at all, or in the (more elusive) sense that the phrase "that elf" (or "that apple"), when directed at a hallucination, picks out—has as its content—nothing at all (despite being meaningful).[31] This general intuition explains not only why we take fictional objects to exist in no sense at all, but also why we take hallucinated or dreamt items to exist in no sense at all.

Before we go on to milk this "general intuition" for ontological implications (about abstracta, in particular), I want to redescribe in what sense determining a criterion for what a discourse commits us to, or determining a criterion for what exists, has proven to be philosophically indeterminate. The indeterminacy is twofold. First, it's not possible to provide reasons for choosing the (first-order) quantifiers as the textual carrier of ontological commitment, versus, say, any of a number of interpreted "existence" predicates. This means, in particular, that when we view our own scientific practice as the generation of *doctrine* (of a certain useful sort), and even when we consider how that doctrine is applied, tested, and generally brought to bear on our world, there is nothing that rationally supports or rules out, say, Quine's criterion for what a discourse commits us to, versus an existence predicate approach that only admits, say, causally efficacious entities, or ontologically independent entities, as existing.

Next, suppose we try to *force* the issue by determining a criterion for what exist—the hope being that the discovery of such a criterion, ontological independence, say, will force us to an existence predicate view (because not *everything* quantified over proves to be ontologically independent of us). This last strategy is dashed because we haven't any (rationally binding) argument for the adoption of a criterion for what exists

---

[30] There is more to say about when to trust (and when *not*) that a usage of "exists" carries ontological force in the vernacular. See chapter 5.

[31] The aficionado will realize that, in trying to indicate what I'm referring to, I'm borrowing a dangerously overused word ("content") from philosophy of language or philosophy of mind. Hopefully, the trained reader can see what I'm getting at through the haze such terminology can arouse.

that rules out a particularly thin interpretation of the existence of fictional entities. So *all* our favorite candidates—ontological independence, causal efficacy, and so on—have an opponent criterion that we can't eliminate. What's left is a more modest proposal—the suggestion that in fact (a sociological fact, if you will) we've (collectively) adopted *ontological independence* as our criterion.

I don't intend to try to characterize "ontological independence" any further in a metaphysical way. But something more can be said about what it means for us to take something to be ontologically dependent, something more that is *epistemic*.[32] If something is ontologically dependent on an author, for example, then he or she isn't seen as needing to square the properties of that thing with something else: There is no requirement that he or she justify the claim that the item in question has the properties attributed to it. If someone isn't making something up, then he or she *is* trying to square its properties with something else and thus *is* required to provide an explanation for why the properties attributed to the item must be the properties it has. This motivates the position that a certain sort of *reliability requirement* on our knowledge claims about items that we take to exist not be susceptible to only a *trivial* explanation:

> *The Reliability Requirement*: The process, by which someone A comes to believe claims about xs, is *reliable* with respect to xs if and only if given that that process has led A to believe $\mathbf{S}x$, then (under a broad range of circumstances) $\mathbf{S}x$, and/or given that this process has led A to believe $\neg \mathbf{S}x$, then (under a broad range of circumstances) $\neg \mathbf{S}x$.[33]

---

[32] Vinueza 2001 attempts a *metaphysical* construal of psychologically dependent items: that they possess only "representational" qualities, qualities held by virtue of being purely psychologically generated items. I prefer to avoid a metaphysical construal of ontologically dependent items and, more specifically, psychologically dependent items, because I presume that we take such items, e.g., the contents of hallucinations or dreams, to exist in no sense at all, and so, therefore, we can't take *them* to have any properties *either*. Still, we might attempt a version of Vinueza's construal by shifting it to a consideration of what sorts of *statements* about hallucinations or dreams we take to be *true*; viz., only those statements that take the form of attributions of representational qualities to the contents of hallucinations or dreams are ones we take to be true of such items. This won't work, however: I can hallucinate pink mice. The (correct) claim that such hallucinated mice are pink is not in the form of an attribution of a purely representational quality (because the property of *being pink* is not a purely representational quality).

[33] A version of this condition first appears as (*) in my 2000c, p. 227. The comments of the next few pages are, for the most part, borrowed and modified from my 2000c. The entire discussion can be impounded to yield conditions on what we take to exist *only because* in *our* community of speakers we take ontologically dependent items to exist in no sense at all.

Talk of reliability may remind readers of Field's (1989) gloss of Benacerraf's challenge to the Platonist, which is to "provide an account of the mechanisms that explain how our beliefs about these remote entities can so well reflect the facts about them" (p. 26). (Also see pp. 27–30 and pp. 230–39.) At the time of my 2000c, I was convinced that the considerations I was there raising, which I encapsulated in (*), were different from those raised

The Trivial Explanation: A process *P* is reliable with respect to *x*s because *x*s have the property that *P* is reliable with respect to them.

Definition: A process *P* is *licensed* as *nontrivially reliable* by an explanation of its reliability with respect to ontologically independent *x*s only if that explanation of *P*'s reliability isn't the trivial one.

The Nontriviality Requirement: If a set of objects are taken to be ontologically independent of us, then we're required to show that *all* our methods for establishing truths about such objects are licensed as nontrivial.

I'll now make several points about these definitions and requirements.

Some may argue that (some) properties of theoretical entities can be introduced by sheer *definition* (Neptune's perturbing of Uranus's orbit comes to mind). If we do so, then surely that these items have the properties attributed to them by definition is both trivial and reliable. This objection suggests that the nontriviality requirement should be weakened to some extent. I'm inclined to doubt this because it can't be that *all* the properties attributed to such objects are established by definition (objects

---

earlier by Field, but now I'm not so sure. The main reason for my thinking the considerations *couldn't* be the same ones was that Field (1989, pp. 28–9) offered a response to Benacerraf's challenge (as he glossed it), which would be an obvious *non sequitor* if his considerations (and Benacerraf's for that matter) were the same as mine. This is that the Platonist could argue according to Field (1989) that

> if mathematics is indispensable to the laws of empirical science, then *if the mathematical facts were different, different empirical consequences could be derived from the same laws of (mathematized) physics*. . . . So, it could be said, mathematical facts make an empirical difference, and maybe this would enable the application-based platonist to argue that our observations of the empirical consequences of physical law are enough to explain the reliability of our mathematical beliefs. (p. 28)

The problem is that the truth of the applied mathematical statements (which is all that the successful application of mathematics can buy us) simply won't provide what's requested: that we have an explanation for why the properties attributed to mathematical objects are the properties they have; and this is because the mere truth of a collection of statements doesn't tell us that the (mathematical) *terms* in those statements hold of the items that the Platonist takes them to hold of (items ontologically independent of us). Or equivalently, the *truth* of such statements doesn't guarantee that they're *true of* the items the Platonist takes them to be true of. See example (3) in the section "Some Illustrations" later.

The same thing happens when Field (1989, pp. 233–8) addresses Lewis's (1986, pp. 111–2) response to Benacerraf's puzzle. Lewis argues that all facts in the mathematical realm hold necessarily. Again, if the reliability concern Field is concerned with is the same as mine, then the appropriate response to Lewis (which Field doesn't make) is that Lewis's response is a *non sequitor*. It doesn't help to describe mathematical statements as necessary, nor will it help even if it's said that mathematical objects have their properties necessarily— say whatever you want to say about *these things*, the issue is how the processes *we use* (including whatever logic we've adopted) to capture mathematical truths is responsive to the properties mathematical objects *have*. This problem is there *whether* mathematical objects have their properties necessarily or not. See example (1) in the section "Some Illustrations" later.

like *that* would be ontologically dependent on us). But then, in practice, what's needed is that the objects have *all* the properties we attribute to them. In such cases, when we learn otherwise, any of the properties initially given by definition are as open to subsequent denial as are later properties attributed to those objects.

This can be overlooked if one thinks of the epistemic process of learning about (certain) ontologically independent objects as proceeding by an initial stipulation of the properties of those objects, and *only then* attempting the satisfaction of an existence claim with respect to that definition. But this is a misleading picture. If all the objects that are candidates for the satisfaction of some definition are ontologically independent of us, then once we've established an epistemic pipeline to (some) object of that sort, we must establish that it has the properties the definition stipulates it to have. Doing this means that our claims about the reliability of that epistemic pipeline must satisfy the nontriviality requirement, so there is no getting past that requirement by means of definition or stipulation or by a process of positing.[34]

The requirement that there be a nontrivial explanation for the reliability of the processes that give us knowledge of objects that exist is a general purpose constraint on knowledge of objects that are ontologically independent of us: The methods leading to our knowledge claims must be ones that we know to genuinely capture the properties of such objects, and not processes that *stipulate* those properties (without any concern for whether the objects will follow suit). In the empirical case, epistemic processes taken to yield knowledge are seen as doing so precisely because they causally connect us to objects in such a way that what the process gives as an answer covaries with the properties the objects have. That is, in the empirical case, the nontriviality requirement is satisfied: that the objects have such and such properties allows a causal process leading from them and their properties to a causal effect on us (via observations, or more indirectly via instrumentation), which in turn contributes to an epistemic process yielding the belief that these objects have such and such properties. But although this is one way of satisfying the nontriviality requirement, nothing in its formulation requires satisfaction in this form; it's stubbornly neutral about what sorts of nontrivial explanations are, in principle, available for the reliability of epistemic processes, and in part, this is because the reliability requirement itself is neutral about how our methods for learning about ontologically independent objects *are* reliable.[35]

---

[34] See my 1999 for a discussion of how Resnik's views, on how the properties of mathematical abstracta are posited, seem to open him to the objection that, contrary to his stated view, such abstracta are not ontologically independent of us.

[35] It's an empirical epistemic fact about us and our universe that certain purported knowledge processes—magical ones, for example—don't exist. Similarly, it's an empirical epistemic fact that we're not Leibnizian monads prearranged to have sensory experiences coordinated with (but not caused by) local environmental events. Were we such Leibnizian

Two qualifications about the reliability requirement: First, the phrase "under a broad range of circumstances" allows an epistemic process (generally) to lead to knowledge even though it may *sometimes* mislead us. For example, under a broad range of circumstances, using our eyes tells us the color of what we're looking at. This doesn't *invariably* yield the right answer, of course: We can be "fooled" by unusual lighting conditions. Second, "and/or" occurs in the formulation of the reliability requirement because, in practice, such epistemic processes may provide only positives, only negatives, or—in very nice cases—both.

We need to know that the methods leading to our knowledge claims about such objects are ones that don't accidentally lead to true descriptions of the properties of those objects. The nontriviality constraint on explanations for the reliability of epistemic processes rules out the accidental reliability of such a process. Imagine that someone lists a sequence of propositions $P_1, \ldots, P_n$ and then flips a coin, stipulating that the coin comes up "heads" on the ith flip if and only if the ith proposition is true. Should coin flips (by accident) match up with all and only the true propositions in this list, then this way of determining the truth or falsity of such propositions is reliable. But only the trivial explanation of why this is so is available.[36]

It needs to be pointed out that, sometimes, the fact that we have only a trivial explanation for the reliability of an epistemic process is *hidden*: A kind of stipulation, or a condition that holds solely by *fiat*, is tucked into the explanation offered for the reliability of that epistemic process (so that it doesn't look trivial at first glance). I give several examples of this sort of case shortly.

The presence of stipulation, although a common indicator that an explanation for the reliability of an epistemic process is trivial, isn't invariably present when only the trivial explanation for the reliability of a particular epistemic process is available. The contents of hallucinations or dreams, as I've already mentioned, aren't stipulated. Nevertheless, there

---

monads, then although the appearance of causality wouldn't correspond to anything metaphysically real, the nontriviality requirement would nonetheless be satisfied. Both the reliability and the nontriviality requirements look like definitional constraints on allowing a process to be knowledge-yielding about objects that are ontologically independent of us. Naturalized epistemology, however, is an empirical subject matter, and so it's an empirical matter *how* those requirements are satisfied. For these reasons, I'm not convinced by Katz's (2002, p. 372) attempt to convict my reliability considerations of presupposing a "contact epistemology," or some form of causation.

[36] There is a nonzero probability, of course, that the coin-flipping process gives the right answer in every case—no matter how many times it's done (the probability of a *run* being entirely veridical diminishes but never vanishes as we let the run increase in length). So, it's *empirically possible* for some epistemic processes that we use to be only accidentally reliable—to admit of only the trivial explanation for their reliability. It's part of the practice of empirical science to try to recognize and rule out such *accidental* reliabilities in our knowledge-gathering activities.

is only a trivial explanation for why we know what visual sensations we're having (say, of a red rubber ball) when in the grip of a hallucination.[37]

I want to rehearse one last point before I use the reliability and nontriviality requirements to provide an argument that mathematical abstracta don't exist. This is that the use of these requirements for metaphysical purposes turns on the examination of how truths about a subject matter are established. These truths, one and all, seem, as a matter of surface grammar, to attribute properties to objects. In deciding that such objects, after all, exist in no sense at all, it isn't that therefore these truths aren't truths or that they're somehow ill-formed, or that their grammatical or logical form is different from what it appears to be. None of these conclusions follow (as I have already shown). All that follows is that, metaphysically speaking, there are no objects of the sort that seem to be talked about, and so, of course, there are no objects with such and such properties.[38]

Here's an argument that we shouldn't take mathematical abstracta to exist any more than we take fictional items to exist: Mathematical abstracta are ontologically dependent on our linguistic practices in just the same way that fictional items are, and because—this is an important part of the claim—the tacit conventions at work in our ontological practices aren't specific to fiction but to *any* collection of purported items that aren't ontologically independent of us, they should be extended to mathematical abstracta.

This argument, however, requires establishing that mathematical abstracta *are* ontologically dependent on us in the appropriate sense. How do we manage that? It's easy. Mathematical objects *can't* be ontologically independent of us because then we wouldn't be justified in claiming that the statements we take to be true of such mathematical objects *are* true. But that we're so justified is simply a datum of mathematical practice (see chapters 1 and 2). Suppose, therefore, that mathematical abstracta were ontologically independent of us. Then the nontriviality requirement on explanations for the dependability of the methods we use to establish mathematical truths would have to be satisfied. But it can't be satisfied in the case of mathematical abstracta.

I can't *quite* show, of course, that the nontriviality requirement can't be satisfied in the case of mathematical abstracta (because this requires

---

[37] This point needs fine-tuning. I can dream about short, stocky creatures and tell people later that I dreamt of elves. It may be pointed out to me, after I describe their visual appearances in detail, that in fact I was dreaming about dwarves. There is therefore a nontrivial explanation available for why we are *right* when we claim (correctly) that we're dreaming about *horses*, say, although nontrivial explanations aren't available for *all* the claims we make (correctly) about such dream creatures.

[38] And here I'm using "there is" in an ontologically committing way. (As the context, I hope, and this footnote, surely, makes clear.) See chapter 5 for a discussion of how we, in the vernacular, indicate ontological commitments.

showing a negative existential claim). What I *can* show is that standard methods of attempting to satisfy it for mathematical objects don't work, and I can give indications for why this isn't a mere technical defect in such methods, but something far more drastic.

Some Illustrations

1. Suppose a philosopher argues thus: Consider a logical system S, and suppose that S generates some truths of mathematics.[39] *Argument*: S is a priori; therefore, the statements about mathematics that follow from S must be true of (certain) mathematical objects. *Response*: What does a priori mean? If it means "Known to be true independently of experience" (however that's cashed out), then S can't be taken to be known in this way to hold of a collection of objects that are ontologically independent of us unless the nontriviality requirement with respect to a priori methods of learning about those mathematical abstracta is satisfied. But what could be said about *us*, and how we've come to adopt S, that could possibly be used to indicate how a priori truths constrain objects that are ontologically independent of us?

Consider, for example, the claim that the a priori nature of S is rooted in the fact that it fixes what's necessary and possible about anything that exists. *Response*: How is that (metaphysical) claim to be established? If we can't *imagine* alternatives to S—even if alternatives to S sound like sheer nonsense to us—that (purely) *psychological* fact doesn't show—all by itself—that the theorems of S are metaphysically binding. For how do we distinguish innate failure of imagination from recognition of metaphysical limitation? Notice that even if we *grant* that there are such metaphysical facts to be found out, we need to explain how *a prioristic* intuitions could possibly be sensitive to such things. No such story even seems *possible*.

Once upon a time such philosophical stories *were* possible and even respectable: Those who believed in God, and in His supernatural might, could (respectably) think that *He* might be so inclined as to construct *us* so that we can recognize such metaphysically binding truths, and recognize *that* they're metaphysically binding. There are, no doubt, questions about how we could prove (a) God's existence and (b) that He really *is* so warmly inclined toward us (esp. given all the *other* evidence to the contrary), but nevertheless, a story seems (in principle) available, *given* these background beliefs. Those of us living in post-Darwinian times, and who think we're at best a pretty smart ape, can easily wonder *what* cognitive mechanism could possibly explain how our intuitions about metaphysical structures are reliable with respect to those structures.

---

[39] Two options: (1) S is a higher-order logical system of some sort with substantial ontological commitments; (2) S is a blend of a weaker logic—such as the first-order predicate calculus—with substantial mathematical or set-theoretic axioms.

Notice that this objection applies not only to views that take mathematical abstracta to be a kind of object that's ontologically independent of us but, nevertheless, something whose properties we can capture by means of ratiocination of some sort; it also applies to views that take the object construal of mathematics to be misguided, to see in the a priori nature of mathematics not knowledge of *objects* but knowledge of conceptual linkages—purely logical knowledge that governs whatever is knowable, or something like that.[40] In this case, we *still* need to supply a nontrivial explanation for the reliability of our capacity to recognize such linkages: We need to know *why* such conceptual linkages *must apply* to what's ontologically independent of us—mathematical and empirical objects, for example. This Kantian-sounding move requires more of Kant's views than is desirable—in particular, his *Idealism*—because only if anything we think about is preconfigured to fit an a priori conceptual structure do we know that the concepts and their linkages that we supposedly generate a priori *must* fit it.[41]

2. This second illustration raises a problem with the applicability of mathematics, given the claim that mathematical objects are ontologically independent of us. Start by noting that it's agreed by pretty much everyone that it's not by causal access to mathematical objects that we learn of their properties. Mathematical objects are taken not to be in space and time, to be acausal, independent of us, eternal, necessarily possessing their properties; and so, access to their properties must occur by some noncausal means. One possibility sometimes raised is that we learn of such objects by examining empirical objects that instantiate them; for example,

---

[40] Such a view assimilates mathematics, or that part of mathematics so assimilatable, anyway, to logic, and grounds our knowledge of the former in the a priori perception of truths of the latter.

[41] Kant (1783) is entirely aware of this, and he's also clear that it motivates his idealism. He writes:

> Sensibility, whose form lies at the foundation of geometry, is that upon which the possibility of outer appearances rests; these, therefore, can never contain anything other than what geometry prescribes to them. It would be completely different if the senses had to represent objects as they are in themselves. *For then it absolutely would not follow from the representation of space, a representation that serves* a priori, *with all the various properties of space, as foundation for the geometry, that all of this, together with what is deduced from it, must be exactly so in nature.* The space of the geometer would be taken for mere fabrication and would be credited with no objective validity, *because it is simply not to be seen how things would have to agree necessarily with the image that we form of them by ourselves and in advance.*" (p. 39, emphasis added)

One concern surely being expressed here is that no explanation seems available for why a priori ratiocination should guarantee that something independent of us must be the way that ratiocination takes it to be. My interpretation: Kant is driven to a (transcendental) nativism because he takes that position to be the only one that can explain why a priori reasoning *must* apply to something; if that something is metaphysically independent of us, no such explanation is possible, but it *is* possible if that something is a product of our own "sensibility."

we learn that 2 + 2 = 4 by examining pairs of objects. On a version of this view, numbers are *predicational* or propertylike. We apply them to, say, collections of objects because they're properties of such collections.

A standard (traditional) objection to the claim that we can acquire knowledge about mathematical objects this way is that this seems to make truths about numbers something to be established by (empirical) induction. I think this objection can be finessed much as Russell (1912) does, *provided* we grant him the assumption that we're acquainted with certain properties via their instantiations in particulars. Given this, although mathematical truths are recognized or elicited by means of empirical experiences, due to our recognition of the necessity of mathematical objects and their properties, we needn't engage in induction or even repeat the experiment (unless we're unsure of what we experienced, a quite different matter).

Instead, I'm going to object that the assumption presupposed by the description of this purported epistemic process, which is that we're taken to somehow *know* which (collections of) empirical objects instantiate which mathematical properties, is unjustified. The point is that if mathematical objects really are ontologically independent of us, we're required to satisfy the nontriviality requirement with respect to our methods of establishing *any* knowledge claims we make about them. This requirement doesn't just apply, therefore, to their purely numerical properties and relations; it must also apply to their relations to empirical objects, including, that is, their *instantiation relations*.

Someone who objects, "Of course we know this sort of thing (how *else* are we, e.g., to count socks, unless we know that a pair of socks instantiates the number 2 and one sock instantiates the number 1?)," has missed the point. The point, after all, is that mathematical objects are presumed to be ontologically independent of *us*, and so we're required to explain how—what's the epistemic process by which—we recognize that a relation holds between an empirical object and something that's located outside space and time, that's acausal, and so on. If the number 2 is a *real* object, which is *really* ontologically independent of us, we need to tell a story about how we *know* that the number 2 has the relations we need to collections of empirical objects, so that it can be used the way we use it *to count*.

It's interesting that the possibility described here, that the number 2 might be instantiated by some couples, some triples, a lonely unit set here and there, barely makes intuitive sense. But what that should tell us is that a picture according to which mathematical objects are ontologically independent of us is one that doesn't square with obvious facts about mathematical practice, in particular, the easy and uncomplicated empirical application of mathematics. We never stop to see if the numbers *really* apply to the items we apply them to, not in the sense that there might be three socks in the drawer instead of two but in the sense—reasonable with respect to objects that really are ontologically independent of us—

that, after all, there might be one sock in the drawer and another sock in the drawer, no other socks in the drawer, and yet, the object 2 not bear the instantiation relation to the socks in the drawer.

To make the issue perfectly plain, here's a somewhat different example: Consider a set of three apples. In this case, what takes the place of the instantiation relation is the membership relation. *What* could be more *obvious* than that each of these apples belongs to the set of these apples? But again, if the set *is* to be independent of us—an item that we *can't* stipulate the properties of, then it's (epistemically (!)) possible for the membership relation *not to be* as it seems to us it *must* be. We *can't* stipulate this, or see it as something that follows from a dictate of "pure reason," unless the set itself is a "conceptual object" of some sort, an item, therefore, that we *can* comfortably dictate the properties of.[42]

Desperate responses are always possible (bumps in rugs can always be moved around by judicious stomping); this shows, someone might claim, that (certain) mathematical objects are located *where their instantiations are* (and so, causal relations, in particular, exist between us and such objects). This *has* been tried, but it doesn't work, primarily because the nontriviality requirement *still* isn't satisfied. When epistemically available causal relations exist between us and ordinary empirical objects, an empirical study can arise of how we make mistakes and how such causal relations give rise to artifacts that mislead us.[43] But no such study exists in this case. We can be mistaken about what properties an object has, but *not* in the sense that the object is exactly as it is (we haven't made a mistake about *it*), but we're wrong about the relations it possesses to abstracta.

Digression

The alert reader has realized that this argument applies to properties *in general*. If properties are (like sets) to be seen as items that are eternal, acausal, not in space and time, ontologically independent of us, but items that are *instantiated* by their instances, we must similarly worry about the satisfaction of the nontriviality requirement with respect to our knowledge of that instantiation relation. Why couldn't we be wrong about something being blue—not in the sense that we think it's blue but we're wrong because of bad lighting, color-blindness, or some other condition

---

[42] With sets, it's almost overwhelmingly tempting for a philosopher to describe the (epistemic) situation not in terms of our perception of an abstraction relation of some sort, but rather in terms of our recognition of *analyticity*, i.e., it analytically follows from the notion of a set that its members just are the items that belong to it. What else is possible? But, then, why describe a set as an *object* that's independent of us, vs. an item that we stipulate the properties of, and so something that is, in a perfectly clear sense, "made up"? Notice the objection: It's not a denial of analyticity; it's that the notion of analyticity is being used here to paper over an epistemic gap by *fiat*.

[43] See chapter 6 for details on this.

(and it's actually red)—but in the sense that the object *is* just as we think it to be, but the instantiation relation between the property of blueness and the object isn't what we think it to be? That such a possibility is so hard to even imagine suggests that properties aren't ontologically independent of us and that, consequently, properties don't exist.

It's not a good answer, no more than in the mathematical case, to treat the instantiation relation as an "internal relation" or an "essential relation" of the object in order to try to stymie the possibility being raised, because the question isn't one about metaphysics but about epistemology: What's it about *us* that gives us an epistemic pipeline to the relations (necessarily or not) possessed by an object that is ontologically independent of us—and so an object we can't *presume* to know the properties and relations of? Some explanation other than the trivial one must be given—that's what the nontriviality requirement requires.

Consider a common explanation: We "abstract" (knowledge of) the property from the object instantiating it. This is only to put a *label* on the problem: *Abstraction* somehow enables our minds to follow up the instantiation relation from concreta to the abstracta they instantiate. But how does it do this? What magic takes us from something in space and time to abstracta that are located nowhere?

Notice that there would be no epistemic problem if this were the view instead: We treat objects that are similar (in some respect) as having (in that respect) a property in common, and the property (of course) is *stipulated* to hold of those objects. But if (contrary to this view) properties *really are* objects ontologically independent of us, then we can't so stipulate *their* properties or relations: Any claim we make about them or their relations is subject to our supplying a nontrivial explanation for why we know these claims.[44]

Staking out a position, as I'd like to, on which properties, too, exist in no sense at all, requires more to be said that can't be said now. So I'll restrict myself to the following remarks. The nonexistence of properties doesn't prevent us from recognizing that objects possess (specific) properties: On this view, we can still see that something is blue, determine that electrons have charge, and so on. Further, we can understand sentences that express such facts to be *objectively* true or false without requiring *objects* of a certain sort—properties—the instantiation relations of which to specific items are needed to *make* such sentences true (or false). That is, it's as objective a fact as one can wish for that snow is white, but the sentence "Snow is white" isn't made true by the existence of an (instantiation) relation between *two* objects, snow and whiteness. Rather,

---

[44] I should stress that, sympathetic as I am to the view I've just described (at least with respect to *some* abstracta), it's *not* my view about properties. I indicate what I think about properties in the next paragraph. Unfortunately, an adequate (full) discussion of properties cannot be undertaken in *this* book.

it's made true by snow's being white, that is, by one kind of thing (snow) being a certain way (white). So the informal use of "possesses" above doesn't betray reference to the presence of a relation, nor does the more tortured locution "being a certain way" betray reference to a kind of thing. *End of digression.*

3. Those familiar with Quine's epistemology and, more generally, with coherentist epistemic positions will consider the possibility that it's systematic considerations about how mathematical theories are designed, and how successful they are in applications, that indicate whether they're true and so whether the posits in such theories exist. Resnik (1997) writes:

> Of course, mathematicians don't even try to detect the objects they posit, since they endow them with no properties that might be detectable. But they do recognize evidence that the theory of the posits is consistent and bears fruitful connections to other mathematical and physical theories as counting in favour of the existence of the posits. . . . [S]uccessful positing is measured in terms of the success of some theory of the posits. . . . (p. 195)

This simply *evades* the demands of the reliability and nontriviality requirements on ontologically independent objects, and this is true, generally, of any coherentist approach of this sort. The problem is that, although establishing the *truth* of mathematical doctrine this way is unobjectable and, indeed, perfectly appropriate (because the role of a mathematical theory with respect to other theories is precisely what tells us whether it's indispensable or not), nothing in this procedure tells us that the *posits* of the theories in question are ontologically independent of us, and consequently, nothing in this procedure tells us whether such posits *exist*.

I've argued in the foregoing that we should be no more committed to the existence of mathematical objects than we are to the existence of fictional ones. But it's important to stress again, despite the ontologically weighty sound of this conclusion, how philosophically lightweight it really is. For I've rooted it in a communitywide tacit convention regarding ontological dependence; that something ontologically dependent on us, the way that dream figures and fictional objects are, is something to be regarded as existing in no sense at all. Were this *not* a tacit convention of ours, as I've indicated, nothing particularly dramatic would change in our epistemic practices. The real issues, in fiction as in mathematics, about how we recognize what's true or false about our claims and, in addition (in the case of mathematics), about how we successfully apply mathematics, remain on either construal of the ontology here. Nevertheless, in part II, I take the stand that mathematical objects exist in no sense at all, and I there turn to the question of what role mathematical posits, in the technical sense of items (1) objectually quantified over in mathematical discourse, although items that (2) exist in no sense at all, play in science.

Contrasts with Schiffer

During the course of chapters 1–3, I've managed myself into a position where I take sentences such as "2 + 2 = 4" or "There are fictional mice that talk" to be *true*, even though I simultaneously claim that numbers and fictional objects *exist in no sense at all*. I've also sketched out a position that takes the same to be true of other abstracta such as properties. One may even suspect that, because sentences are types, and propositions are most commonly taken to be abstracta, that I probably don't think they exist *either*. But Schiffer (1996) argues that such objects are "mind and language independent in at least two senses" (p. 150). First, such items, such as the property of being a dog, exist in possible worlds "in which there are neither speakers nor thinkers, which is just to say that something might have had, or failed to have, the property of being a dog even if there had been neither thinkers nor speakers"; for that matter, such items exist—in this sense—in the past of the *actual* world, in which, presumably, there were neither speakers nor thinkers. Second, such properties and other items "belong to no language."[45]

These reasons play as much a role in motivating Schiffer (1996) away from a view that takes properties (and fictional objects) to exist in no sense at all as does his presupposition that objectual quantifiers must be ontologically committing, and that no other sort of quantifier (e.g., substitutional) will do the anaphoric job that the objectual quantifier is needed for. What can I say to assuage such concerns?

I'll take the "belong to no language" concern first, because it can be dispatched quickly. The worry is that if such items really were language dependent, they would be dependent on specific languages. And yet they don't seem to be. This is a genuine concern only for those philosophers who take abstracta to be real entities that are also language dependent. The nominalist, who takes them to not exist in any sense at all, has only the burden of explaining why it's convenient to design identity conditions across different languages for noun phrases (and predicates) that (actually) refer to nothing at all.[46]

Regarding the first concern, let's, for argument's sake, take the metaphysical situation regarding possible worlds and properties to be as Schiffer (1996) describes it. Because the "cash value" of the existence of properties, for example, amounts to their holding or not holding of things in

---

[45] Schiffer (1996) argues that the same is true of propositions: "To say that the proposition that snow is white exists in possible worlds in which there are neither thinkers nor speakers is merely to say that the proposition would have existed (and had a truth-value) even if there had been neither thinkers nor speakers" (p. 166, n. 5). Fictional objects become mind independent on his view via the claim, which he endorses, that they are abstract objects.

[46] For an example of how this can be done in the case of mathematical languages, see my 1994, pt. II, §§ 6–7. The terms of natural languages don't pose any further special problems.

those worlds, and because, in turn, I take this to amount to the truth or falsity of sentences or propositions describing such states of affairs, and further, because the truth or falsity of sentences or propositions describing such states of affairs, I've argued, is independent of the ontological requirement that properties exist, there is no problem. Such statements are true or false of possible worlds just as they're true or false of the actual world, and considerations about truth and falsity of sentences in possible worlds don't require the existence of properties (or other abstracta) any more than do considerations about truth and falsity in the actual world.[47] Statements about abstracta, for example, "2 + 2 = 4," are even easier to handle: We simply stipulate that such statements have the same truth values in possible worlds as they do in *ours*.

But the reader may still suspect that a problem is lurking here: Can *sentences* (or propositions) be treated as true or false, if they're abstracta and so, as far as the nominalist is concerned, don't exist either? Well, if the nominalist *does* want to claim that sentences or propositions *don't* exist, there *isn't* a problem in doing so for reasons already given in chapter 3: To say that certain sentences are true (or false), and to supply (say) a semantics that takes such sentences to be true (or false), requires (if we take the Tarskian route) a metalanguage in which quantifiers range over a domain rich enough to satisfy the semantic needs of a theory of truth. Such (objectual) quantifiers needn't commit us to the existence of abstracta—sentences, propositions, mathematical objects of various sorts, and so on—any more so than do any other objectual quantifiers.

But, in any case, despite *my* nominalism, *I* wouldn't want to claim that (every) sentence (or proposition) exists in no sense at all. As I discuss in part II, my general view is that we must distinguish the posits in *pure* mathematics from the posits in *applied* mathematics. The posits of pure mathematics exist in no sense at all, and if the pure mathematics in question is the mathematics of formal languages, for example, then the notion of a *sentence* or of a *proposition*, as such notions arise there, indicates classes of posits of pure mathematics—these are abstracta that exist in no sense at all. However, applied mathematics is a different matter because (often) in application (some) of the mathematical posits of mathematical doctrine so applied *proxy* for other things that *do* exist. My view is that the entire semantic apparatus for truth (when applied to an actual language) is itself a branch of applied mathematics, and so sentences (or propositions) in such a context can proxy for something that does exist but that, within the context of semantics, is described via abstracta (which exist in no sense at all). Similar things happen in physics. Exactly *how* mathematical posits proxy for what's real will be the focus of part II.

---

[47] I hope my language here won't mislead: "States of affairs" is merely a picturesque way to generally refer to things like snow being white, glass being transparent, etc. States of affairs don't exist any more than properties and relations do.

Concluding Remarks about Criteria for What Exists

I've argued that it's a metaphysical given in our community of speakers that objects that are ontologically dependent on our linguistic processes, such as fictional characters or mathematical objects, and objects that are ontologically dependent on our psychological processes, such as the contents of hallucinations or dreams, don't exist. Notice that this isn't a condition on what it's reasonable for us to take to exist or a condition on what we're justified in taking to exist. We're justified in taking the contents of our hallucinations not to exist *because* we take it as a condition on what *exists* that it not be dependent on psychological processes the way that the contents of hallucinations are. However, ontological independence *of us* in this sense isn't yet a *criterion* for what exists because it does *not* supply necessary *and* sufficient conditions on what exists, but only a *necessary* condition.[48]

On the other hand, the reliability and nontriviality requirements are necessary and sufficient conditions—so I've argued—for us to take posits *of ours* to exist. This is because a posit that doesn't fulfill such conditions fails to be ontologically independent of us (as far as we know), and that's a *necessary* condition on what exists, as I've just mentioned. It may seem, therefore, that a criterion for what exists (necessary *and* sufficient conditions for what exists) isn't needed. We are, in any case, only concerned with our own posits and how to evaluate them metaphysically, and a necessary condition on what exists seems to suffice.

This last sentence isn't true, however, because we take the range of our quantifiers—regardless of whether our theories are true or not—to include everything that exists.[49] Because we want a special existence *predicate* that holds of all and only the items that exist, and this predicate is to function the same way that other ordinary ones do (e.g., "gold"), it must hold of what exists and fail to hold of what doesn't exist. If ontological independence of us is only a necessary condition on existence, it can't be used to underwrite the semantics of such an existence predicate—only a condition that is necessary *and* sufficient for existence can do that.

However, I underplayed my hand when I described us as taking ontological independence (in the required sense) of *our* psychological and linguistic processes to be what we, as a community, take as a condition on

---

[48] Recall the discussion of necessary and sufficient conditions on existence earlier in this chapter. There may be *terms*, i.e., that are nondenoting and so, when appended to the existence predicate, yield falsehoods, despite it being a truth that appending such terms to "ontologically independent of *us*" yields a *truth*.

[49] I've argued, of course, that they range over *more* than everything that exists—at least in one respectable way of understanding "range over"—but the argument I'm giving only requires that they at least range over everything that exists.

I should add that it's possible for *ontological independence* to outstrip our domain of discourse and infect other notations in our language, i.e., predicates. For the most part, I've avoided exploring the ramifications of this possibility; I will, however, raise it explicitly for certain polemical purposes in chapters 8 and 9.

what exists. In point of fact, we require that such posits be independent of any psychological or linguistic process *whatsoever*—whether ours or someone else's.[50] The latter *does* supply necessary and sufficient conditions for what exists. I call such posits *ontologically independent* (not just ontologically independent of *us*).

I close this chapter with several points about this last gloss on ontological independence: First, this definition of ontological independence is acceptable even for those who believe in a supernatural designer of the universe. There is work, of course, needed to spell out the precise differences between "ontological independence" in the sense where the universe is dependent on its designer in the way that an artifact is ontologically dependent on its designer, and the very different sense, which I'm using, where the "universe" is dependent on its designer because He is only engaged in reverie (and actually hasn't brought anything into existence). I leave this particular task for another time and place (and, possibly, for another philosopher).

Second, that we take ontological independence in this sense so seriously explains the widespread intuition that the "things" the evil Cartesian demon depicts to its victim, or what the brains-in-a-vat experience, aren't "real."

Third, as the first observation indicates, I've not done a lot toward spelling out exactly what's involved in something being "independent of psychological or linguistic processes." That's OK—these are empirical notions, as the words "linguistic" and "psychological" indicate—and thus the notion of ontological independence is subject to the same sorts of refinements and corrections that *any* empirical notion faces (e.g., *virus*). There's no reason why the conditions governing an existence predicate need be more precise than other empirical notions, or any more immune to revision than are the conditions associated with other (empirical) predicates.

Fourth (and finally), to repeat, there's a sense in which "ontological independence" doesn't describe an actual *property* of what exists because it doesn't oppose this property explicitly to a contrary *property*, "ontological dependence," which all and only things that do *not* exist have. This is because, as I've said before, what does *not* exist exists in *no sense at all*— what doesn't exist doesn't have *any* properties. What's *really* happening here, as I described earlier, is that *terms* of our theories are being distinguished, and the ways that we characterize them as picking out something that exists (or not) are also being distinguished.

---

[50] Compare this with Moore's (1939) careful discussion of the notions "to be met with in space" and "external to our minds."

5

Ontological Commitment and the Vernacular: Some Warnings

Adopting the nonexistence view toward fictional and mathematical objects (which I've urged for in chapter 4) forces us away from (when regimenting ordinary language in first-order logic) the strategy of reading off ontological commitments from the quantifiers. Instead, one or more regimented predicates must be taken to carry such commitment.

Once Quine's criterion, and his triviality argument for it, is gone, we've *lost* a comparative ontological tool; we no longer have a purely syntactic criterion for recognizing what *others* are committed to. As a result, the ontological evaluation of discourse becomes a great deal more difficult: Given so many possibilities for how ontological commitment *may* be expressed in the vernacular, and so many ways it can, in turn, be regimented formally, we must engage with ordinary (domestic *and* alien) discourse more subtly than a first-order existential quantifier criterion requires in order to recognize how genuine ontological commitments are best regimented.[1] And our own ontological commitments, and our own ways of recognizing such commitments within our own discourse, don't supply any sure guidance when it comes to alien discourse. This is an unavoidable loss.

Early on, in chapter 3, I (qualifiedly) attributed a claim to Quine that, in any case, I'm committed to: Ordinary speakers make—consciously and tacitly—genuine ontological claims. Supporting this view may seem to

---

[1] I illustrate this claim several paragraphs later.

require a fairly thorough linguistic study of the vernacular: the attempt to locate, say, a specific predicate, or perhaps a special *kind* of quantifier expression, which—in the vernacular—unequivocally carries ontological commitment. One might *even* hope that predicates in English such as "exists" and its cognates have such a role.

Indeed, there are reasons to think that perhaps "exists" *does* have such a role, and consequently, there are reasons to think that this bears against the attempt in chapter 4 to assimilate the ontological status of mathematical abstracta to that of fictional entities.

> Opposing Consideration: Unlike the case of "there is," which is an idiom that really does crop up in discourse regardless of whether we take the subject matter to which it's applied to exist in any sense at all, we *really are* loathe to apply "exists" in certain contexts.

Although we're perfectly willing to say, in the vernacular, that "prime numbers greater than 7 exist," we're not willing to say, "Fictional mice that talk exist." So why not claim that ordinary people *are* willing to commit themselves to numbers, that they're *not* willing to commit themselves to fictional entities, and that "exists" really is an ontologically committing predicate *in the vernacular*? We should consider this line of thought very carefully. Although it may be correct, we shouldn't be motivated (as the objection insinuates) to treat fictional entities and mathematical entities—ontologically speaking—in different ways.

First off, the Opposing Consideration needs refinement because the linguistic evidence for an ontological role for "exists" (in contrast to "there is") isn't as straightforward as that Consideration makes it appear.[2] Van Inwagen (2000, p. 239) points out that we do sometimes attribute "existence" to fictional items. Here are his (very natural-sounding) examples: "To hear some people talk, you would think that all Dickens's working-class characters were comic grotesques; although such characters certainly exist, there are fewer of them than is commonly supposed." And: "Sarah just ignores those characters that don't fit her theory of fiction. She persists in writing as if Anna Karenina, Tristram Shandy, and Mrs. Dalloway simply didn't exist."

We can avoid being forced to the choice of either denying that "exists" operates ontologically in the vernacular, or denying that we—collectively—take fictional objects to exist in no sense at all, by arguing that "exists" *does* (very often) operate as a locally contrastive device *without*

---

[2] One area of discourse that can't be used as evidence either for or against an ontological role in the vernacular for the word "exists" is fictional discourse itself (vs. discourse *about* fiction and its fictional entities). This is because even though sentences like "Although gremlins, goblins, and orcs do exist, dragons, alas, do not" occur *in* stories, they fail to indicate anything about the word "exists." On the pretend-true view of fictional discourse itself (to which I subscribe), the ontological force, if any, of uses of the word "exists" in fiction is nullified; this maneuver isn't available for mathematical discourse (by contrast) because (applied) mathematical statements must be taken as true.

ontological significance, but that this isn't the case when no contrast is drawn. An indication of this difference in use is our reluctance to say *straight out*: "Fictional mice that talk exist" (because it sounds like an ontological commitment to fictional mice), although there is no resistance to saying "There are fictional mice that talk."[3] This suggests that "exists"—used noncontrastively in the vernacular—carries ontological force.

We must be careful about this claim, however, both with respect to the supposed contrast with "there is" and with respect to the claim that we can this easily distinguish, in the vernacular, noncommitting uses of "exists" from committing ones. There is the point I mentioned in chapter 4 (n. 19): "Really exists" doesn't sound redundant to the untutored ear. (It doesn't sound bizarre for someone to say "Numbers don't *really* exist.") This implies that rhetorically unstressed uses of "exists" need *not* carry ontological force, and that ordinary speakers know this. The point is that tone and stress play a big role—both with "there is" and with "exists"—to indicate to speakers whether or not what's being talked about is something that one takes oneself to be committed to.[4]

Another point is that, although it's true that "Mickey Mouse exists" simply sounds false, it doesn't sound false to say "Mickey Mouse exists in the world of Disney." Thus, in discourse *about* fiction, it's not uncommon to talk about what exists and what doesn't exist in the fictional world of some author. But surely "in the world of So-and-so" doesn't (semantically) cancel the ontological force of "exists," if it has any. (Lewis's view, that there *really are* possible worlds in every bit the same sense that there

---

[3] It's important, when harvesting intuitions about examples like these, to be clear about the (implicit) contexts that are and aren't being presupposed when imagining the utterances of sentences. The stand-alone claim "Fictional mice that talk exist" *will* sound natural if previous sentences (perhaps by other speakers) provide implicit contrasts. So, e.g., imagine someone making claims about what sorts of fictional animals there are, and ignorantly claiming that although there are lots of talking rats, there are no talking mice. This might be opposed by the utterance of "Fictional mice that talk *do* exist." This is different from those contexts—which *exist*—where speakers are engaged in a straight-out discussion of ontology. (The reader should *not* assume that cases where speakers are engaged in straight-out discussions of ontology are always philosophical ones where "language is on holiday." Arguments about whether God exists or not, for example, are ones that ordinary people, *even extremely unphilosophical ones*, understand the implications of *very well*, and are *very* concerned about.)

[4] Thus, the tone and stress used with "There are fictional mice," plus context, can indicate ontological commitment to such items. And with both idioms "there is" and "exists," speakers have the option of being unequivocal via rhetorical enhancers: "There *really are* fictional mice" and "Mice *really* exist." This suggests that the intuitive difference in ontological commitment between unstressed stand-alone uses of "There are fictional mice" and those of "Fictional mice exist" may be due to subliminal pressure from the unconscious inclusion of particular contexts of utterances in the one set of cases (and not in the other), and therefore isn't a symptom that it's part of the semantics of "exists" (vs. "there is") that it carry ontological force.

is an actual world, may be false, but this is hardly to be established by the ontic-canceling power of "in a possible world," when affixed to "exists.")[5]

If the foregoing remarks on "exists" are right, they imply that the search for symptoms of ontological commitment in the vernacular via a search for *specific* locutions with *specific* ontological roles is the wrong way to go. For if I'm right, ontological indicators involve, inextricably, pragmatic devices such as rhetorical stress, coupled with ways that we have, in ordinary discourse, of deliberately stepping out of that discourse and commenting upon it. ("Of course there aren't any prime numbers, *really*, because, *really*, despite how mathematicians talk, numbers don't exist" or "Numbers don't exist in any sense *at all*.")[6] Just as the Quinean regimentalist was chasing a will-o'-the-wisp in hoping to fix on technical grounds a logical device with ontological significance, so too, I'm willing to suggest, semanticists studying the properties of natural languages pursue a will-o'-the-wisp if they hope the study of the logical form of the vernacular will betray semantic or syntactic devices that (unequivocally) indicate ontological commitment. In *this* sense, not to say that there aren't others, ontological commitment *really* belongs to the domain of philosophy *proper*.[7]

It's easy to see why ordinary discourse is unlikely to have idioms specifically designed *only* to carry ontological freight, and they're similar to the reasons why first-order objectual quantifiers needn't do so. The anaphoric needs that the objectual quantifier fulfills, and the semantic needs that predicates and singular terms satisfy, are all essential to the expression of truths about a subject matter *regardless* of whether what's spoken of exists in any sense at all. So, too, the more immediate need to

---

[5] I'd argue that this is even true if we think "exists in the world of Disney" is short for "exists in the *fictional* world of Disney." This is because the only cogent way I know of to make this run are along the lines of the cancellation option discussed in chapter 3, and for similar reasons, it won't work.

[6] In this book, for example, I've repeatedly availed myself of rhetorical intensifiers in order not to be misunderstood; I've repeatedly said, for example, that fictional objects exist *in no sense at all*.

[7] In the foregoing, I've been concerned with what—*beyond* the empirical objects of science and ordinary life—we might take ourselves to be committed to (to take as "existing"). But as the history of philosophy—esp. ancient philosophy—makes clear, there are other options. Imagine a neo-Platonist who takes "unchangingness" to be criterial of what exists. Such a view is *compatible* with how "exists" works in the ordinary vernacular. Such a person will say that ordinary physical objects don't (really) exist; only abstracta do. A certain style of philosopher—call him an "ordinary language" philosopher—might see neo-Platonism (so construed) as a typical example of how (certain) philosophers try to abuse ordinary language without paying for it. My counterclaim is that certain idioms—"exists" in particular—are *susceptible* to such "abuses" precisely because of how they *are* used in the vernacular.

I'm *not* claiming historical accuracy for the details of this example; I suspect *Plato*, anyhow, derived *unchangeableness* as a criterion for what exists from something more basic (i.e., the noncontradictoriness of property attributions to what exists).

make *local* ontological contrasts, plus the opportunistic tendency of language users (over time) to warp language tools to handle "nearby" tasks,[8] explains why it's unlikely that there are idioms in ordinary language dedicated solely to marking a general distinction between what *really* exists and what *really* doesn't exist. Finally, it's likely that within a community of speakers, it's generally unnecessary to be explicit about one's ontological commitments. When the ordinary speaker says, "There are fictional mice that talk," he or she doesn't even think to try to avoid (by rhetorical stress) the suggestion of a genuine ontological commitment to fictional mice because it's very rare that an explicit concern with ontology will arise in any case.[9]

Some may think that the localized contrastive role of "exists" and "there is" shows that something like a Carnapian view of ontology is the right one: Separate languages—even language games—are operative in our ontologically fragmented body of beliefs, and all have their specialized internal "exists" predicate, or specialized quantifiers. More general claims about ontology are meaningless "external questions." Leaving aside how confirmation holism makes such a mess of this sort of view, reasons are already in place for why this isn't a good position to take: It's clear that ordinary people have ways of making ontological claims that are more supple and elusive than what predicates and quantifiers need indicate on a semantic basis alone (and this, as I've said, is because the quantifier idioms and specialized predicates, e.g. "exists" and its cognates, have been co-opted for the purpose of adjudicating divisions *within* subject matters—regardless of the ontological status of such subject matters). Ordinary speakers can (fairly easily) be brought to understand general questions of ontology, however: This isn't a matter of becoming confused about how "there is" and "exists" operate semantically—it's an illustration of the ordinary person's ability to recognize a question the expression of which isn't rigidly codified via specific idioms.

Given the truth of my claims about quantificational idioms, and specialized predicates, such as "exists," however, it's easy to understand the Carnapian temptation to deny the bigger ontological question any sig-

---

[8] No linguistic device, no *word*, for that matter, is guaranteed to remain fixed in its usage. Speakers, from generation to generation, are quite willing to put idioms to new uses that can obliterate the possibility of a straightforward expression of what was previously meant by that idiom. The history of word usage abounds with examples, and with writers who are willing to be morally outraged at the loss of a particular easily expressed nuance.

[9] Such a concern does arise when there's a need to be explicit because one is in a philosophical context, or because one is dealing with an individual with very different ontological commitments: I stress again, because of how *ordinary* they are, conversations between the irreligious and the religious. This also contributes to an explanation for why, in the vernacular, the idioms for making ontological commitments explicit are in thrall to what seem to be *obvious* pragmatic devices: Ontological issues come up rarely *not* because ordinary people have no real understanding of what such issues are about, but because the ontological commitments of one's discourse are usually understood in exactly the same way by like-minded folk—family and friends.

nificance at all.¹⁰ Still, we philosophers can get at what's wanted when speaking in the vernacular in pretty ordinary ways (using rhetorical enhancers and by explicitly stepping out from a way of speaking and commenting upon it), but when we regiment ordinary discourse, we can happily *desert* these logically dubious tools of ordinary language and instead capture the ontological commitments (in the most general sense of this phrase) of users by *designating* a predicate (say) of the regimented discourse as doing so. That such a regimented predicate won't correspond to *all* uses of any single locution in ordinary language is no objection to the cogency of this procedure, or to the meaningfulness of what's being designated. Indeed, such a state of affairs supports the continuing need *to* regiment discourse: The subtlety of ordinary language with respect to ontological claims is all too easy for philosophers (intent on generalization) to misunderstand. To regiment is to put such matters plainly for all to see, and—there is no getting past this—it's also to make our ontological *decision making* equally plain for all to see.

The potential treachery of ordinary language, with respect to how it indicates the ontological commitments of ordinary folk, doesn't arise *only* with idioms, such as "exists" or "there is," ones seemingly concerned with ontology. Various sorts of epistemic idioms, for example, offer similar traps. Consider the word "see." Boolos (1998) declares himself "rather a fan of abstract objects, and confident of their existence" (p. 128). His reasons turn on the fact that we immediately see these things, and he forcefully writes:

> It would be a rather demented philosopher who would think, "Strictly speaking, you can't see *The Globe* [a Boston newspaper]. You can't even see an issue of *The Globe*. All you can really see, really immediately perceive, is a copy of some issue of some morning's *Globe*." To say this, however, reflects a misunderstanding of our word "see": more than a misunderstanding, really, it's a kind of lunacy to think that sound scientific philosophy demands that we think that we see ink-tracks but not words, i.e. word-types. (p. 128)

He goes on, a page later, to note that we "deal with abstract objects *all the time*." We listen to *radio programs*, we write *reviews* of *books*, we correct *mistakes*, and some of us draw *triangles* in the sand and write *numbers* on chalkboards. To think otherwise about numbers *on* chalkboards, Boolos suggests, is to be confused about the use of the word "on."¹¹

It seems to me entirely clear that Boolos is quite right about ordinary language. "See," for example, enjoys a straightforward usage extending all the way from fuzzy colors on surfaces to baseballs and bats to neutrinos in the sun to various types and other abstracta. And surely (as an embarrassingly large amount of awkward epistemological doctrine from

---

¹⁰ I succumbed to such a temptation, and pretty much for these reasons, at the end of my 1998. Also, compare Austin 1962, pp. 62–77, and Bennett 1969.

¹¹ I hear echoes of Austin here. See Austin 1962, p. 31.

Bishop Berkeley on shows), there is *no* escaping this by restricting what one "sees" in some precise sense to what's "immediately seen" or what's "really seen." But anyone would be rash to draw ontological conclusions *of any sort* about the wealth of items ordinary language licenses us as able to see, because innocently included among such items are hallucinations: "I'm seeing elves again," the aging hippie in the grip of an acid flashback will say. "I must have had too much to drink," someone else (under slightly different circumstances) will add, "because there are those pesky pink elephants again. I'm *so* tired of seeing those things all the time." If we can *see* more than is dreamt of in the philosophies of nominalists, that's only because we can see more than there is.

Concluding Remarks

Having said all this, I hasten to add that I wear my newly adopted nominalistic garb lightly. This isn't only because I hold that nominalism is a position that can't be adopted for philosophically convincing reasons (but only—at best—on the grounds that the most general intuitions prompting folk ontology tend toward nominalism), but because even with respect to the latter considerations, there is still a question of *exactly which* promptings of barely conscious folk-ontological intuition we should be swayed *by*. Raising this issue brings us back to the Opposing Consideration—specifically, to the suggestion that ordinary folk ontologically distinguish mathematical posits from their fictional kin. I'm now going to set aside my qualms about whether "exists" carries ontological commitments in contrast to "there is," and argue that even if this is so, and even if it's clear because of this difference in the usage of "exists" that ordinary people take fictional entities to exist in no sense at all but do take mathematical abstracta to exist, we (philosophers) shouldn't follow suit, despite the fact that there are no definitive philosophical reasons for taking an ontological stand about this, one way or the other.

I've claimed that *ontological independence*, in the appropriate sense, should be the decisive consideration on which the existence of something should be decided, and I've suggested that folk ontology is with me on this one.[12] But the Opposing Consideration suggests that my claim is debatable: that the reason that it intuitively feels more natural to describe mathematical abstracta as existing than it does to so describe Mickey Mouse is because the dark promptings of folk intuition treat mathematical abstracta differently from how they treat fictional entities. I concede this.

---

[12] This *doesn't* exclude, of course, other properties being empirically co-extensive* with ontological independence in the sense that what we take to be ontologically independent we also find to have those properties. E.g., it could be (and I think it is—see part II) that items that are ontologically independent in the required sense are also causally efficacious (and vice versa). My only claim is that, as a matter of folk intuition, the notion of ontological independence appears to be more directly linked to the notion of existence than is the notion of causal efficacy. (This is indicated by the belief held by many philosophers that mathematical objects are ontologically independent, causally neutral, *and* that they exist.)

Further, folk intuition, I suggest, is motivated by differences between fiction and mathematics: There seem to be objective questions of right and wrong in the case of mathematics, and it seems that statements in mathematics have truth-values regardless of whether there is any method, even in principle, for determining what these are; neither point seems true in fiction, and, related to this, there seem to be discoveries to be made in mathematics versus the inventions (the "fancies") that writers of fiction routinely generate. These impressions of differences sway untutored intuition away from the suggestion that mathematical abstracta exist in no sense at all and toward a picture of abstracta where they're the *source* of the purported objectivity of mathematics because *they* are the items that mathematicians discover the properties of.

What moves me to urge resistance to *these* promptings of folk intuition that impel a belief in mathematical objects is that they not only impel the view that mathematical abstracta exist but also motivate a specific take on the properties of such abstracta. The picture suggested by untutored intuition is just that centuries-old Platonic picture of items that are eternal, not in space and time, the truths about which are necessary, and so on. But *this* picture, unlike the bare claim that mathematical abstracta exist, really *is* incoherent when examined in relation to respectable epistemology, because such items must be, apart from existing, ontologically independent of us, and this (I've argued) mathematical abstracta can't be.

The right picture for existing mathematical abstracta, if one must have them, is the one corresponding to the second fictional realist view: items brought into existence at just the moment we pen references to them on paper (or think them up), and which have the properties they have only because we take them to have such properties. And, as I've argued, there are no definitive philosophical arguments against someone who takes *this* view of mathematical abstracta. But those of us who'd like an ontology that both adequately corresponds to our body of beliefs *and* doesn't repeatedly give rise to philosophical misunderstandings would do better to urge *everyone* to adopt the view that mathematical objects—like the fictional entities they so closely resemble—exist in no sense at all. Doing so is to engage in regimentation in a stronger sense than the mere adoption of a regimented predicate to carry ontological force even though (perhaps) no such predicate unequivocally exists in natural language: It is, in the name of clarity, to resist the temptation folk ontology offers (at least in some of its moods) of treating mathematical abstracta differently from fictional entities.

One last point: My nominalism, at least as far as applied mathematics is concerned, is compatible with what has come to be called "semantic realism."[13] This is the acceptance of classical logic with the entailment—mentioned a few paragraphs earlier—that every sentence (in the lan-

---

[13] This take on realism originates with Dummett and has been subsequently adopted by Quine, Putnam, and others. See, e.g., Dummett 1963 and Quine 1981b.

guage(s) of applied mathematics) is either true or false independently of whether there is any method, even in principle, of determining such truth-values. The acceptance of classical logic for applied mathematical doctrine is due to the need for a topic-neutral logic governing both empirical claims and the mathematical doctrine applied to those claims, coupled with the already-in-place adoption of classical logic for empirical propositions. Because fiction isn't applied in this sense, it doesn't need to be governed by the same logic that governs the rest of our body of beliefs (applied mathematics, the empirical sciences, and ordinary folk doctrine).[14]

Luckily, a commitment to semantic realism doesn't bring with it metaphysical requirements on what sorts of things, if any, the semantic realist must be committed to. This is simply because the mere adoption of classical logic doesn't necessitate any conditions on how subsequent determinations of undecided mathematical statements are to be made, *if ever*. So semantic realism isn't incompatible with the practice of augmenting mathematical systems with additional axioms and stipulating co-referentiality* between terms appearing in such systems for reasons totally internal to the practice of mathematics.[15] In light of this, I should say that there are reasons to think that the term "semantic *realism*" is a misnomer.

---

[14] It's likely that there is no "logic" of fiction—at least no *uniform* logic of fiction. This isn't true, of course, for discourse *about* fiction, or about fictional characters, because those claims are part of our ordinary body of beliefs.

[15] E.g., the elegance and ease of provable theorems, the usefulness of the augmented system in applications, the surprising ways that results can be established, etc. The point of the term "internal" is that considerations of how to square mathematical claims with ontologically independent items are not included.

Further details on how the view of mathematics I'm committed to is compatible with semantic realism may be found in my 1994, esp. pp. 140–9, 210–4.

II

APPLIED MATHEMATICS AND ITS POSITS

Posited objects can be real.

W. V. Quine

6

Posits and the Epistemic Burdens They Bear

I aim, in the rest of this book, to examine a small number of small-scale scientific theories and evaluate, in the light of the ontological apparatus constructed in part I, their ontological commitments. But before examining these theories in any detail, some preliminary work has to be done first: To this end, in this chapter, I initially sort the posits of theories into several categories based on their epistemic characteristics, and then in chapter 7 I draw ontological conclusions about these posits by applying arguments from part I. (The reader will see from this process, if he or she doesn't know it already, that there is nothing straightforward about moving from epistemological doctrine to metaphysical claims.)

Let's begin: *Posits* are the purported referents of singular terms (names and definite descriptions) wherever such terms arise in our discourse: ordinary life, the sciences, mathematics, discourse about fiction, and so on. So too, and more important from the scientific point of view, among our *posits* are the items picked out by quantifier claims, remarks to the effect that "There are Ss," even when particular Ss can't be singled out by either name or definite description. "Posit" is a term borrowed from Quine, and it's a strange term, as he uses it, because it simultaneously nods to the world (the posits of our theories exist) and to theory construction (posits are items we "posit" when we adopt theories). Thus, "posit" seems stylistically tainted with a certain use/mention conflation: What's attributed to posits might sometimes be better attributed to the *terms* purporting to refer to such things, and as commonly used, "posit" also seems to involve diffidence about commitment (in ordinary speech, we only "posit" what we don't *know* to be true). Quine entirely disavows the latter contrast when he uses the word—at least with respect to those posits *we* take to exist: We *know* those to exist as much as we know *anything* we know—and harmlessly glories in the first conflation.[1]

The Quinean can formally distill out the distinct ways posits nod to the world and to our theories by focusing on differences between the quantifiers and the predicates. On Quine's view, our quantifiers range over what exists, not over what we merely take to exist. This is why our theories can be *wrong*. (If our theories only concerned what we took to exist, then those theories would not be falsified by what actually exists and how it deviates from what we take to exist.) On my modification of this view, the quantifiers are no longer *restricted to* what exists, but still, the falsification of theories turns (in part) on what there is—which the range of the quantifiers includes—not being as those theories claim it to be.

For the Quinean, this is compatible with another sense in which the "posits" of our theories are only what we *take* to exist. We take frogs to exist—this requires the predicate "frog" to be instantiated, and it may not be, but this is a very different point from the one made in the preceding paragraph, and notice how it turns on the instantiation (or not) of a *predicate*.

In describing this (Quinean) view about predicates, I'm *not* agreeing that the role of predicate and quantifier crisply divide in this way. They *do* on the Quinean view, I think, because the extensions of predicates are determined *solely* by the theories governing predicates. But on my view, the predicate—if it's a "criterion-transcendent term" (or a "natural kind term" as it's more commonly put)—is independent of theory insofar as we may *change* the theory and yet regard the predicate as continuing to pick out what it always picked out (the theory we took to govern the predicate was wrong). (See my 2000b, pts. III and IV, for details about how I think this goes.)

Hereafter, "posit" is a technical term. Although (1) it "refers to" the purported references of individual noun phrases or to the items that common nouns hold of, and (2) it's ontologically neutral, because what a noun phrase picks out or holds of may not be taken to exist in any sense at all (something, as I indicated in part I, that I believe to be true of mathematical abstracta and fictional characters), nevertheless, when a discourse is regimented, (3) it always corresponds to the results of *quantifier commitment*, the derivability of sentences of the form $(\exists x)Sx$, where Sx is any formula with variable x free.

Notice that Quine's criterion determines what those committed to a discourse (ours or someone else's) *take* to exist—hence the theory relativity ("derivability") of the existential implications. It's a different matter what *exists* on this view (which existential statements are *true*), but, of course, these *explicitly* separate only with respect to *alien* discourses.

---

[1] For the most part, Quine's choice of terminology hasn't misled. See, however, Collins 1998.

For Quine, and those who adopt his criterion for what a discourse is committed to, (2) and (3) contradict each other. But the nominalist position I've taken toward mathematical abstracta in part I requires my dropping Quine's criterion for an existence predicate ("ontologically independent") instead. Nevertheless, it's stylistically harmless to speak of posits even if they're taken to exist in no sense at all, and so for smoothness of exposition I'll do so. In the following, I distinguish (1) *quantifier commitments*, those commitments of a theory that are due to its existentially prefixed implications, and that tell us what posits a theory has, from (2) *ontological commitments*, those commitments that we take the theory to have because of those of its posits that, in addition, are ontologically independent.[2]

It's important to realize that the (epistemic) distinctions I draw later between types of posits are independent of any particular ontological position that may be taken about those types. As I've already indicated, one can disagree with me about whether mathematical abstracta exist or not and yet (1) agree entirely about the epistemic distinctions among posits that I draw and (2) agree entirely about where mathematical abstracta (and fictional items) fall given those distinctions. This methodological insight about how the ontological and epistemic aspects of my position *don't* stand and fall together is facilitated by a choice of terminology ("posit") that I treat as ontologically neutral.

I distinguish posits by whether they have *epistemic burdens* and, if so, what kinds. Two points about such burdens: First, in calling them *epistemic* burdens, what I intend to stress is that success in satisfying them provides a prima facie reason to think that the posits exist (and, indeed, provides the best prima facie reasons *we have* for thinking such things exist). Second, such epistemic burdens can't be translated into "analytic entailments" that hold of the *general terms* designating the posits. This is because such burdens are defeasible: It's always open to us to change the epistemic classification of a posit (and, indeed, the classification of *all* the posits falling under a term) and thus to change whether it has epistemic burdens or not (and which kinds).

Let's first consider those posits that have no epistemic burdens at all. *Ultrathin* posits are what I call such items, and they can be found—although not exclusively—in pure mathematics where sheer postulation reigns: A mathematical subject with its accompanying posits can be created ex nihilo by simply writing down a set of axioms;[3] notice that both individuals and collections of individuals can be posited in this way.

---

[2] As I indicated at the outset of this chapter, I'll largely keep to the viewpoint that the ontological commitments of a theory are to be found *among* its quantifier commitments. Later, however, I'll consider relaxing this condition on the grounds that what a theory should be taken to be committed to—because of the ontological independence of the things in question—may be found among other linguistic paraphernalia of a theory, i.e., its predicates.

[3] See my 1994, esp. pt. II. Ultrathin posits, of course, include fictional entities.

Sheer postulation (in *practice*) is restricted by one other factor: Mathematicians must find the resulting mathematics "interesting." But nothing else seems required of posits as they arise in pure mathematics; they're not even required to pay their way via applications to already established mathematical theories or to one or another branch of empirical science.

This aspect of mathematical practice poses a problem for those who prefer to think that mathematical posits exist and are ontologically independent, because our commitment to such posits—such as it is—seems strangely resistant to change. Mathematical theories can always be neglected, but in being neglected, it's not that we've changed the ontological status of the posits in such theories (by, say, contrasting their current fall from grace with a previous state of *Being* that we attributed to them before that neglect set in), because should subsequent mathematicians find some neglected branch of mathematics interesting, they can research the area, and this *won't* be described as research on chimeras—as it would be if a contemporary physicist tried to study caloric fluid or the ether (or if a biologist tried to study *chimeras*, for that matter—I mean by "chimera" the goat/snake/lion amalgam of ancient Greek mythology). Neglected mathematical posits are taken to be exactly to the same extent that more popular mathematical posits are; it's just that no one (else) cares about them.

This lightweight epistemic status of ultrathin posits contrasts with the epistemic burdens that *thin* posits have. *One* of the requirements on these is, in Quine's words (1955, p. 247), to provide the theoretical virtues of simplicity, familiarity, scope, fecundity, and success under testing. That is to say, individuals introduced into a theory by name or description, and collections of such introduced via predicates that hold of them, are to be governed by *statements* (a theory) that exemplify Quinean virtues. I call these epistemic obligations <u>Quinean rent</u>. Notice that Quinean rent can be owed by a single posit (picked out by a definite description or name that plays a role in a theory) or by a collection of such (where a predicate plays a role in a theory).

In my 1994, I coined the term "thin posit" to describe the sort of posit that only has living up to Quine's five virtues as its epistemic burden. I'm dropping adherence to that terminology in this book, and here describe *thin posits* as posits required both to pay Quinean rent and to fulfill a certain defeasibility condition (for why we *haven't* thick epistemic access to them). Examples of thin posits (in my current sense) are items such as any frog that no human has seen (or otherwise detected via instrumentation), various viruses (most of them) that haven't been specifically detected, and so on. More details about thin posits follow the amplifications of the conditions on thick epistemic access.[4]

---

[4] The reader may wonder if there are *any* posits that are thin in my earlier 1994 sense—that only have Quinean rent as an epistemic burden. I don't think so, as I argue later—and I'm partially indebted for my views on this to a discussion with Aaron Lipeles back in 1995 or thereabouts.

The difference between thin posits and ultrathin posits (which live free of charge) is striking. Should one of the former fail to pay its Quinean rent when due, should an alternative theory with different posits do better at simplicity, familiarity, scope, fecundity, and success under testing, then we have a reason to deny that the thin posits, which are wedded to the earlier theory, exist—thus, the eviction of centaurs, caloric fluid, ether, and their ilk from the universe.[5]

Thick posits labor under a different epistemic burden: They're items to which we have "thick epistemic access."[6] An individual's (or group's) thick epistemic access to a posit can be purely sensory in nature (someone looking at something or listening to it), can involve individuals and instruments (microscopes, radar), or can involve a complicated social network of individuals and instruments (e.g., computers) interacting (as a group) with something, where there is a division of labor (e.g., the operation of a linear accelerator).[7] Regardless, there seem to be four conditions on thick epistemic access:

(1) The results of thick epistemic access to something are largely independent (epistemically speaking) of what the recipient(s) expects from that access (I call this *robustness*).[8]
(2) There are means of adjusting and refining thick epistemic access to the things being detected (*refinement*).
(3) Thick epistemic access to things enables tracking of them (either in the sense of detecting what they do over time or in the sense of taking time to explore different aspects of them) (*monitoring*).
(4) (Certain) Properties of the objects can be used to explain how the kind of thick epistemic access we have to them enables the discovery of (possibly other) properties of those objects (*grounding*).

Amplifications

In what follows, I comment on conditions (1) – (4) of thick epistemic access by considering details about the two kinds of thick epistemic access: on the one hand, ordinary observation by the use of our naked senses, and on the other, instrumental interventions with items otherwise outside of our sensory range.

---

[5] The reader may suspect that centaurs were evicted simply because we didn't find any to begin with. That's usually not quite enough of a reason to rule something out if it's an ontological belief we've inherited, although it's a strong consideration. Additional theory is often needed to convince us that our failure to find something isn't merely bad epistemic luck. Here is where the Quinean virtues come into the picture.

[6] See my 1997b and 2000b, p. 116. Much of the following discussion on thick epistemic access is drawn (and modified) from those sources.

[7] See, for nice historical examples, Galison and Hevley 1992.

[8] "Robustness" is a term that often comes up in philosophy of science; one shouldn't assume that it means here what it means elsewhere.

On Robustness

The independence in this sense that *robustness* supplies includes—but isn't restricted to—independence from one's *theories* about what one is thickly accessing, as well as independence from one's theories about how that thick access itself operates.[9] When thick access is a matter of pure observation, the independence of its results from what we believe or expect arises from the nature of our ordinary evidential practices with respect to observation. (To illustrate: If I want to prove the existence of something to someone skeptical of it—e.g., moldy bread in the refrigerator—and it's possible to point at it, I can do so immediately.) This independence is nevertheless compatible with the following facts: We can be wrong about what we think we see; we can be trained to see things we couldn't see before; we may need a bit of theory in order to understand what it is we're supposed to be seeing, or how seriously we should take what we're seeing. And finally, there is no denying that there can be states of mind—great emotion, drugs, mental illness—where the independence of our sensory access to things from our beliefs about those things is compromised.

When thick epistemic access operates via scientific instruments of some sort, its robustness is (in part) due to the antecedent robustness of observation. This is because the use of such instruments always involves transduction of the machinations of what we're thickly in touch with into something we *can* observe. If a particular instrumental intervention with something regularly leads to a certain result, this means that any of a certain class of ordinary observations occurs when the device is operated. For example, a common tool for recognizing the presence of certain acids (in certain forms) is litmus paper, and dipping a bit of it into a liquid can result in a change of color that we can see. So, too, more sophisticated devices may yield flashing lights, certain sorts of visual images, sounds, and so on. There is no requirement, however, that any particular sort of instrumental intervention *must* be accompanied by a particular sort of observation (a Geiger counter makes clicks; something else that measures the same thing may produce a photograph); I'm only saying that when an instrumental interaction is robust, it achieves this (in part) via a link to observation.

But this link to ordinary observation isn't the only source of the robustness of (instrumental) thick epistemic access to something. A source of robustness, in *all* cases of such access, is an unavoidable limitation on *confirmation holism*; that is, it's due to our inability to bring scientific theory directly to bear upon the events involved in thick epistemic access, whether that access be a matter of pure observation or a matter of instru-

---

[9] I must stress that this independence is never *total*. Theory can, often in unpredictable ways, have an impact on our understanding of our thick epistemic access to something. The point is that, as a matter of empirical fact, such impact is fairly limited.

mental intervention—this means that we can't use scientific theory to predict what will happen when we gain thick epistemic access to something or to circumscribe when we can trust what we seem to have learned via that access (about something).

One reason we can't bring scientific doctrine to directly bear on the thick epistemic access relation is the sheer complexity of the macro-events involved: macro-beings—ourselves—are *always* at one end of that relation. Thus, in having to empirically learn to navigate successfully (via our senses or with instrumental tools) without the detailed use of scientific doctrine, we must acquire a cruder body of empirically derived regularities instead. This is done empirically by practicing with those tools (or with our senses); many of these learned regularities are kinesthetic, that is, a matter of instinctively manipulating the instruments (or ourselves) in certain ways.[10] I call these "gross regularities," and they're what underlie our capacity to execute the methods of evidence-gathering that we use to gather knowledge empirically. As a result, these evidence-gathering methods, both in ordinary life, and in the sciences, remain (for the most part) epistemically independent of the sciences precisely because the gross regularities necessary to them can't be deduced from science itself.

There are both *observational* gross regularities, which underpin our capacity to observe, and higher-level gross regularities, which can be discovered about theoretical entities—such as electrons—and which enable us to study the latter's properties in ways relatively independent of our theories about those items. In particular, gross regularities are the source of both *refinement* and *monitoring* in thick epistemic access.[11]

On Refinement

When it comes to ordinary observation, we're quite familiar with the many ways that we can (and do) instinctively adjust and refine our thick epistemic access to something: We can squint to see things more clearly; we can move in for a closer look; we can reach out and touch. Thus, we have resources to help distinguish what we're "really" seeing from what's only an artifact of our senses (or of how those senses track items in the world). How do you tell that "floaters" are in your eyes and not several inches in front of you? By engaging in various experiments that enable you to determine that the motions of such items are due to the move-

---

[10] These regularities are cruder than the truths of pure scientific theory in at least two respects: First, they're often limited in scope—applicable when instruments are used in certain specific situations but inapplicable, even with the same instruments, in other (sometimes only slightly) different situations. Second, they're usually deductively independent of *other* such regularities; i.e., they're usually established empirically on an individual basis, without there being much of a hope that they can be connected (deductively) to other regularities or, for that matter, to anything else.

[11] See my 2000b, pts. I and II, for extensive discussion of gross regularities, in particular, for details on their epistemic relation to scientific doctrine, and for examples.

ments of your eyes. Notice that these ways of refining, adjusting, and understanding our sensory access are robust: They're (largely) independent of theories held about why such methods of refinement work.

Instrumental access to theoretical entities is entirely analogous: One reason for refining our instrumental access to theoretical objects is the achievement of robustness; the tinkering needed for this often goes on independently of our theories about the objects accessed or of the theories about the instruments used (we simply "fiddle" to reduce "noise" of various sorts). But, as with pure observation, crucial to this process of refinement is also the recognition of artifacts of the instrumental process itself, and here theory can play—although it doesn't always—an important role. What we see under the microscope might be an artifact of how we dyed the sample and what effect that has on light. We may recognize this either because we know (on the basis of a theory) that the item in question doesn't have the structures we seem to be seeing, or because we can isolate this artifact by using other instrumental means of access to the object under scrutiny. Because of the epistemic need to detect and eliminate artifacts in our means of access to something, a significant amount of experimental design is dedicated to the elimination of such artifacts.[12]

On Monitoring

One nice property of thick epistemic access is how it enables its recipient to "track" the properties of something. For example, you can watch something for several hours (a certain sort of insect, say), and in this way determine a variety of its properties (that it moves at a certain speed, that it can change colors, that it likes moss). Furthermore, you can determine what it does over time; you can construct, that is, an "episodic history" of its actions in real time.

Instrumental access to theoretical entities allows the same thing. Historically, in fact, the capacity to track a theoretical entity instrumentally

---

[12] A classic discussion of this with respect to the microscope is in Hacking 1981 (reprinted in his 1983). Two nice case studies where concern with the identification of artifacts is also explicit are Franklin 1993 and Weiskrantz 1986. Also see my 2000b, pt. I, esp. § 6, and also pp. 175–6. It's worth pointing out that this sort of study doesn't just take place with respect to thick epistemic access to the micro-objects physicists, say, are concerned with. The psychological literature, for example, often raises the worry about whether one or another psychological notion has been sufficiently or correctly *operationalized*. What's innocently indicated in these contexts by the use of this (philosophically scary) word is only the concern with whether the experiments in question have established a rigorous enough connection, means of access, between the experimenters and the phenomenon they hope to study; or, in other words, whether what they trust they're measuring is, in fact, being measured. As an illustration (there are many), see Bills 1981 on worries about the self-concept, and Cunningham and Tomer 1990, p. 384, on fluid intelligence.

Similar worries arise in social psychology and sociology—when the phenomena in question are macro-items that one must get to by means of statistical tools; consequently, statistical artifacts are a notoriously ever-present danger.

seems to have been an important consideration if scientists were to come to a consensus on its existence. Back in 1913, Jean Baptiste Perrin (1990) included as part of his important evidence for the existence of atoms C.T.R. Wilson's experiments that seemed to show the actual trajectory of atoms through an atmosphere saturated with water vapor. So, too, we find the following evidential considerations that "made the 'particle' identification of quarks more secure" (Adair 1991, p. 1002):

> Measurements of the scattering of very high-energy electrons, accelerated by the Stanford Linear Accelerator (SLAC), from neutrons and protons demonstrated that the nucleons contained, or were made of, point-like charged particles which scattered the electrons strongly. Through analyses of the magnitude of the scattering, the partons were identified as the fractionally charged quarks. (p. 1002)[13]

On Grounding

We can often connect certain properties of the objects we can see with our ability to learn about their properties through sight; for example, given our (partial) understanding of our observational means of access to something, and given what properties we take the seen object to have, we can understand how we're able to recognize, by sight, its color, speed, and so on. This need not be a subtle scientific matter (although it often is): It's easy to explain why I can see how fast a flock of antelopes is moving—they're large opaque objects that don't travel very fast (even when panicked). So, too, it's easy to explain my troubles seeing how big certain jellyfish are under certain circumstances: They're (more or less) transparent; but, consequently, I know I won't have a problem seeing how big they are if I manage to dye them first. Thick epistemic access to theoretical items otherwise impervious to sensory detection is entirely analogous on this point.

The four conditions on thick epistemic access aren't all on a par as far as their epistemic status is concerned. *Grounding* is different from the other three conditions because it depends more directly on empirical assumptions about how thick epistemic access works. Imagine, for example, that the old Leibnizian view is right: Causation isn't metaphysically real; instead, there is a divine coordination of perception of event with event perceived. Even in such theologically loaded situations, thick epistemic access would be just as epistemically useful as it is for those of us here under more Godless circumstances. *Grounding*, however, would lapse:

---

[13] Also see Brown et al. 1997. In that article, we find: "The third jet in these 'three-jet events' is now recognized to be due to the emergence of an energetic gluon in a process analogous to *bremsstrahlung* in QED. Although it took a few more years to make an absolutely convincing case, this *visual* evidence for gluons was perhaps the most influential factor in the acceptance of QCD as the correct theory of the strong interactions" (p. 21, emphasis original).

the detail-oriented scientific explanations (of how *this* specific property of *that* enables us to track it because of certain causal interventions we're consequently capable of) would be replaced by a general all-purpose global coordinating explanation involving design on a cosmic scale by a benign divinity.

This *doesn't* mean that the other three conditions aren't susceptible to *empirical* explanations for why they have the epistemic value they have. *Robustness*, for example, is due to the empirical fact that simple and easily applied theories that are also predictively adequate and that are computable in *real time* (i.e., quickly) aren't to be had. *Refinement* and *monitoring* arise from details about the properties of what there is (e.g., that things are spread out in space and time in certain ways) and details about how our sensory capacities (augmented by instrumentation) can intersect with how those properties manifest themselves in space and time to enable us to learn about things. Of course, the manifestation of what there is is never a pure matter of the properties of something simply being "given" to us. In this sense, therefore, the epistemic centrality of observation (and, more generally, that of thick epistemic access) isn't a fundamental *value* that's to be assumed by empiricists without argument: It can be seen to arise from what there is and how we manage to know what we know about it.

This explains why we can (to some extent) imagine possible worlds and beings in them where observation, in particular, would not have the value for those beings that it has for us. For example, imagine that simple and easily applied theories that are predictively adequate, and computable in real time, *were* available. And imagine, further, that observation is, for such beings, the same sloppy tool that it is for us. Under such circumstances, they might *not bother* to observe: As a means of access to *knowledge*, theory derivation would simply be more efficient, less work, and more accurate.[14]

One point: In saying that we can tell a story of what we and our world are like that uses scientific descriptions and that explains why observation and, more generally, thick epistemic access have the value they have, I'm *not* claiming that it's our capacity to tell such a story that *confers* epistemic value on observation and thick epistemic access. That these epistemic methods have such value *already* (and regardless) is embedded in our evidential practices—and these are practices that we can't give up. All that's shown by the foregoing discussion of how scientific studies of how the structure of our world (and us) enables us to learn what we learn about the world is that our picture of the world (and us) *coheres with* the inalienable evidence-gathering practices that we have and, in particular, coheres with the general usefulness of such practices for finding out what's true.

---

[14] One thing that makes this hard—but not impossible—to imagine is that the possible world depicted (not to mention the beings in it) has to be one that is itself a great deal simpler than our world is.

And what if these aspects of our body of beliefs didn't cohere? There are three possibilities. First, we could find ourselves *changing* our epistemic practices (successfully). This has happened: The (real) possibility of generalizing observation to thick epistemic access is something we discovered only around the beginning of the twentieth century. Second, we could find ourselves *changing* our view of ourselves and the world. This has happened, as well—as the history of optics from the ancient Greeks to modern times indicates. Third, and least appealingly, we could acquiesce to the inconsistency.

"Inconsistency" is perhaps too strong a word. One aspect of the *discomfort* with standard interpretations of quantum mechanics is that measurement has to be interpreted within classical physics despite the fundamental physical theory (of the small) being nonclassical. This radical discontinuity ("collapse of the wave function") seems a given of quantum mechanics as traditionally viewed, and this explains the dualistic mind/body feel of, for example, the Copenhagen interpretation, and may also explain the tendency of some physicists—and popular expositors of quantum mechanics, as well—to sense compatibility between quantum mechanics and certain "spiritualistic" views. Nevertheless, there is clearly a strong desire not to acquiesce to this state of affairs if we don't have to. And, views *have* emerged that seem capable of explaining the classical physics of macro-objects from within a purely quantum-mechanical perspective. If any of these are successful, this will make our body of beliefs coherent (in the broadly logical sense), because then the same physical laws will govern both our naturalized epistemology and what there is at what we (currently) take to be the most fundamental level.

Given that *grounding* does hold of thick epistemic access, one important implication is that items to which we have thick epistemic access are ones the properties and relations of which play a role in the epistemic stories we tell about how we know about them and what means of thick epistemic access we can have to them. I call this epistemic obligation of thick posits their *epistemic role obligation*.[15]

The required epistemic stories, about how instrumental thick epistemic access from us to posits is possible, are ones told by scientists (in various subdisciplines) about how their instruments, and their other means of interactions with the world, function. So, for example, there's the story physicists tell about how the instruments they've developed to measure acoustical phenomena operate.[16] There's also the story that psychologists

---

[15] As just shown, any claim that any empirical posit whatsoever *must* satisfy an epistemic role obligation must turn on *grounding* always holding of thick epistemic access. In any case, it's not true that satisfying the reliability and nontriviality conditions of chapter 4 *requires* a fulfillment of the epistemic role obligation.

[16] In Strasberg 1991, we find an explicit discussion of a number of measurement tools (various kinds of microphones and hydrophones, the Rayleigh disk, fiber optics, etc.), their range, sensitivity, and other pertinent properties, in the specific context of physical theory. I.e., a certain group of causal relations between us and certain phenomena (in this case, fluctuations of pressure, temperature, density, etc., of matter) are studied.

borrow from physiologists about how our visual systems work. These stories are invariably incomplete, as stories often are in the sciences, but the important point is that they're *epistemic* stories—they're clearly designed to indicate how we know what we know about something—how, as far as the science of a particular time can tell us, the means of access that we have to things operates.[17] For this reason, the study of empirical phenomena, and thus empiricism as a practicing method among scientists, isn't correctly described by merely saying observation (or, more generally, thick epistemic access) is crucial. Rather, it's that a systematic study is also required of the means of access to empirical objects, along with concomitant worries about artifacts of that access.[18]

Because posits can be both thick and thin, posits sort into four types: Ultrathin posits, thick but not thin posits, thin but not thick posits, and those that are both thick and thin.

Thick but Not Thin

Individual posits usually play no role in theories, and as a result, they rarely pay Quinean rent. But they still have epistemic role obligations. If I can't explain how I know a particular turtle, Peter, is in my garden, or if I insist on talking about my crystal ball when people ask me how I know about Peter, these are pretty good reasons for thinking Peter doesn't exist. The same is true even when thick epistemic access is forged to theoretical objects or collections of such: A stream of electrons to which we have thick epistemic access is still a collection of (specific) items that play no (significant) role in theories; that *they* exist offers little by way of Quinean virtue. Still, we must tell a story about our instruments that explains how we gained access to *them* in particular.

Thin but Not Thick

On the other hand, any predicate that we take to hold of items *some* of which we have thick access to we generally also take to hold of (sometimes numerous) items to which we don't have thick epistemic access.

---

[17] What makes it especially clear that these stories are *epistemic* stories is precisely the focus on correcting mistakes mentioned in the section titled On Refinement above. Notice that nothing of this sort exists when it comes to the purported epistemic access to abstracta of various kinds. This, on the one hand, illustrates how thick epistemic access *is* sensitive to the requirements of the reliability and nontriviality conditions of chapter 4 and, on the other, puts in a stark light how purported epistemic access to abstracta is *not*. I give more details about this shortly.

[18] Neither observation nor thick epistemic access (which generalizes upon observation) is a metaphysical black box that achieves its epistemic deeds by magic. They're both as transparent to scientific inquiry as any other aspect of our universe.

"Turtle," for example, we take to refer to *every* turtle whether it has been sighted by anyone or not. "Electron" we take to pick out electrons everywhere, regardless of whether they're the (relatively few groups of) items we've forged thick epistemic access to, the slightly larger number of electrons we could have forged thick epistemic access to but didn't, or the numerous electrons that we've no hope of forging thick epistemic access to. These posits are *thin*.[19]

Thick and Thin

Positing individuals, of course, is sometimes crucial to a scientific theory, and then the individual posits must pay Quinean rent. Think of Neptune: Positing Neptune was simpler (easier, surely) than other changes in physics we might have considered in light of Uranus's erratic orbit—this much is Quinean rent—but nevertheless we had to gain thick epistemic access to the planet itself before we'd believe in it.

All the standard empirical objects *studied* in one or another scientific discipline—chairs, cats, ants, neutrons, cells, viruses, anger, multiple-personality syndrome, nation-states, and so on—are posits, some (or all) of which we take ourselves to have thick epistemic access to and the rest of which we take to be thin posits.

Two Points

First, thick epistemic access is a *relation* between a community and objects during a time period. This means that not only are new instances of the thick epistemic access relation forged between us and objects to which we didn't earlier have thick epistemic access, but also that the thick epistemic relation to items can be lost later (e.g., Roman artifacts that we discover later in time, other items—speeding atoms, rabbits bounding in the brush—that leave the narrow domain within which we can observe them or otherwise instrumentally detect them).

Second, aren't *all* the items *referred to* by "turtle" either thick or thin posits? No: We can be wrong about what we have thick epistemic access to—we think those things are turtles, but they're not. Second, we can be wrong about thin posits, too: We think there are no turtles on Pluto, for example, but we could be wrong about that. More generally, we could be wrong about the theory of something we use for determining that those somethings exist (out of our ken).

---

[19] "Turtle," of course, refers to turtles, whether we take them to exist or not. But the thin posits that are turtles are those turtles that we take to exist on the basis of our theories about turtles even though we *haven't* got thick epistemic access to them. *Are* there such turtles? Presumably; see the discussion starting several paragraphs from now.

Those philosophers who want to use observation (or its generalization to thick epistemic access) as a criterion for what we take to exist[20] would draw a very strong moral from the Neptune example I gave earlier: They would argue that we won't let ourselves be committed to the existence of something unless we have thick epistemic access to it. But this is clearly false, as the turtles just mentioned illustrate. We take it that there *are* turtles that we (collectively) haven't got thick epistemic access to.

Still, something in this initial idea can be saved, so it would seem, because even though we haven't got thick epistemic access to *every* turtle we take to exist, it isn't simply that these other turtles that we take to exist are merely obligated to pay Quinean rent; it isn't that we merely say that we believe in such turtles because of simplicity, familiarity, scope, fecundity, and success under testing. In addition, a certain *defeasibility condition* is required: We have an (implicit) story in place about why these turtles we take to exist nevertheless are items we don't have thick epistemic access to. It's not a very difficult story to tell: We haven't explored (or destroyed) every last inch of the terrain where turtles are apt to hang out. In general, the range of items we can forge thick epistemic access to is fairly limited—and so it's easy to see why items in other parts of space and time aren't things we've forged thick epistemic access to.

That is to say, not only do thin posits have the Quinean virtues as an epistemic burden, but they also have the epistemic role obligation, *as well*. It's easy to see why: Turtles seen and turtles unseen have the same properties and obey the same laws. Thus, the application of *grounding* to turtles we have thick epistemic access to, which gives rise to the epistemic role obligation, also applies to unseen turtles (the general reasons we give for why we have seen such and such turtles can be used to explain why *there are* other turtles we haven't seen.) But in the case of unseen turtles, the epistemic role obligation calls for a defeasibility condition (an explanation for why we *haven't* seen these turtles).

With these considerations in place, those philosophers I mentioned earlier who want to use thick epistemic access—in some form—as a criterion for what we take to exist can suggest that we don't take thin posits to exist unless they fall under a kind-term some of the instances of which are items we have thick epistemic access to (for it's only by such posits falling under such a common kind-term that *grounding*, which provides information on items we have thick epistemic access to, can be transformed into a defeasibility condition for those posits). And these philosophers can remind us that the history I rehearsed earlier about atoms seems to indicate this: We didn't become convinced of the existence of atoms of any sort until thick epistemic access was forged to some of them. Then we accepted the existence of *other* atoms, regardless of whether we had thick epistemic access to them or not. The same seems true of biological

---

[20] A criterion for what we *take* to exist, not, I stress, a criterion for what exists.

types: We discover a new kind of frog, and then we're convinced of the existence of other frogs of the same kind.[21]

But this suggestion, too, seems open to a potential counterexample. It's taken to be a certainty that there are numerous *kinds* of viruses to which we haven't yet forged thick epistemic access. (Indeed—and try not to be *too* upset by this—new kinds of viruses are quietly mutating on this very page as *you* read.) The same is true of *kinds* of insects, plants, and even mammals: We take it to be *certain* that there are kinds of insects, plants, and even mammals no instance of which we've yet forged thick epistemic access to.[22]

Why do we think this? There are two reasons. The first involves theory: We know, on the basis of theory, how variable viruses, in particular, can be, how fast they can evolve, and how widespread they are. Similarly, we know how evolution works and that isolated species restricted to local environments can easily develop, and that such unexplored environments, although vanishing rapidly (e.g., the Amazon rainforests), are still with us. The second is empirical: We read, almost on a daily basis, about discoveries of new species of animal and plant. A similar point can be made about chemical compounds: We know—without thick epistemic access to them—that there are *kinds* of chemicals that we haven't discovered yet. So we're firmly committed to the existence of items, even to the existence of items of specific kinds, where we *don't* have thick epistemic access to (any of) the instances of those kinds.

The key to understanding what's going on here is to recognize, first, that the notions of *kind* and *subkind* implicitly at work in our positing thin posits are ones generated by theory.[23] Consequently, it's theory that tells us, when faced with a kind, examples of which we've thick epistemic access to, whether we know enough about that kind (*and* its subkinds) to take ourselves to know that there are thin posits falling under that kind (and its subkinds). In the case of viruses, theory licenses the belief not only that are there specific (individual) viruses that we haven't thick epistemic access to, but also that there are *kinds* of viruses that we haven't thick epistemic access to either. On the other hand, until relatively recently, despite our acquaintance with the planets in our solar system, we simply didn't have enough theory to tell us that there are other planets ellipsing around other stars to which we lacked thick epistemic access.

---

[21] Although not *always*—and this shows how theoretical considerations are used to convince us (or not) of the existence of thin posits. If this frog has very specific needs and can, as a result, flourish only in very specific habitats (which, say, have been eliminated by one or another oil company), we may not believe there are any other frogs of this type to be found. We may take him, tragically, to be the last of his kind.

[22] Could we be wrong about this? Of course—we can be wrong about anything. But the point is that we take ourselves to *know* that there are kinds of things that exist no instance of which we have thick epistemic access to, just as we take ourselves to know, e.g., that there are electrons.

[23] See my 2000b, esp. pts. III and IV.

Interestingly, in the case of planets, this isn't because we hadn't enough theory—in principle—that tells us whether planets are likely to be common or not; it's that the theory we had was too intractable to apply directly to cosmological data (to get *these* answers). Recently, however, empirical results enabled us to verify the existence of planets around other stars, and our discoveries that such planets exist under quite varied circumstances has licensed the claim that there are many planets to which we haven't thick epistemic access.

Causation and Thick Epistemic Access

Although all thick epistemic access is causal, not all causal contact with things is thick epistemic access to them. One reason is that awareness that one is in epistemic contact with something is a requirement on one's having thick epistemic access to that thing. Someone struck on the head by a falling object may not have thick epistemic access to that object simply because he doesn't "know what hit him."[24]

It's easy to think, however, that the previous necessary condition, of the awareness of the recipient of thick epistemic access to something, places stronger constraints on thick epistemic access than the example I've given implies. Thick epistemic access to something, one may think, turns on background knowledge about that thing: If I'm struck by a lightning bolt, it's my knowledge that lightning is a stream of electrons that enables me to have thick epistemic access to that stream of electrons. Without this knowledge, I've thick epistemic access to a yellow bolt of something or other that hurt me (Zeus's thunderbolt, for example).

This is the wrong way to view thick epistemic access. In particular, regardless of whether I know that thunderbolts are *streams* of electrons or not, I've thick epistemic access to one when I see it, although I don't have thick epistemic access, of course, to any of the individual electrons that make it up.

What *is* true is that thick epistemic access to something is a matter of *our* causal relations to that thing: Whether we have thick epistemic access to something turns on the causal relations we have to that thing that enable thick epistemic access to be forged. That is, how an item that a recipient of thick epistemic access is interfacing with is individuated by that recipient turns on exactly what sorts of causal relations he or she has to what's being interacted with; it doesn't turn on what background knowledge the recipient has about what he or she is interacting with. A baby, for example, (eventually) has thick epistemic access to all the objects around her that she can sense: Nevertheless, she doesn't (at least for a while) know what most of what's she's interacting with (and learning

---

[24] More generally, those causal effects taking place within us and to us that we're unconscious of, because, say, their effects are too small for us to perceive, are ones to which we don't have thick epistemic access—at least not through introspection or perception.

about) is. Similarly, that my senses aren't fine-tuned to a certain degree allows my thick epistemic access to masses of water and prevents my thick epistemic access to individual water molecules, and this is regardless of whether I in fact know what water is composed of.[25]

I don't want to give the impression that thick epistemic access—or observation, for that matter—requires us to be able to distinguish (neatly) every single item in a group of somethings in order to be described as having thick epistemic access to those items as *individuals*. We can observe a herd of antelopes and distinguish clearly that they're antelopes, even if we can't quite count how many there are or track them (individually) well enough to give each one a name. And there may not be a sharp distinction between this sort of case and that of the water molecules (imagine, as an intermediate case, a clump of jellyfish out for a swim)—observation, after all, is an empirical relation, and precise conditions on how well we're able to individuate items in order to claim that we can *see* them may well prove elusive.

This brings us to a somewhat treacherous terminological matter. I've described thick epistemic access as a generalization of observation, and I've also, when discussing some remarks of Boolos's in chapter 5, mentioned that we can *see* more than there is. These remarks, if coupled (as I've just done), make it clear that "observation," when described as obeying *grounding*, can't be the notion of "observation" where we're allowed to "see" objects that don't exist in any sense at all.

And that's right: The sense of "observation" that is "generalized" upon by thick epistemic access is a "factive" notion. Regardless of whether there really is a use of "observe" or "see" in ordinary language that is entirely factive (something I doubt, actually), I'm introducing such a notion of "observation" for epistemic purposes. In this (technical) sense of "observe," we can't observe what doesn't exist because seeing such "things" prevents satisfaction of *grounding*. So, too, should we turn out to be wrong about what we took ourselves to be observing or otherwise instrumentally interacting with (Bs, say), the appropriate thing to say is that we were *wrong* about the presence of thick epistemic access to Bs; this is compatible, by the way, with claims that we actually had thick epistemic access to Cs (some other kind of object entirely) or that we had no thick epistemic access relation to anything at all.

---

[25] At least, I don't have thick epistemic access to water molecules via my unaided senses. The point of scientific instrumentation is to change our causal relations to the world and, as a result, to change what we can have thick epistemic access to.

I realize that what I've claimed, about how what we have thick epistemic access to is individuated, is controversial: Among other things, it implies that the various indeterminacy doctrines held by many philosophers are false—at least insofar as those doctrines require that the causal relations we forge to the world are insufficient to distinguish, say, the different individuation conditions associated with "rabbit" and "rabbit-fusion." See my 2000b, pt. III, §§ 2–3, or my 2003, where further argument is offered on behalf of causation and how it enables us to individuate what we interact with.

The amplification on *robustness* described how limitations on confirmation holism and the consequent existence of gross regularities allow thick epistemic access to be (relatively) independent of theory. But I should say a few more things about how theory operates with respect to thick epistemic access (when it does) before concluding this chapter.

The important point is that theory operates with respect to thick epistemic access *descriptively*, not *constitutively*. What I mean by this can be illustrated by something already discussed: I've stressed how (a specific) theory may be needed to sort out those aspects of (a particular kind of) thick epistemic access to something that reveal actual properties of the object versus merely exemplifying artifacts of the means of access itself. In playing this role, theory doesn't create the distinction between *artifact of means of access* and *property actually held by something* any more than the theory of subatomic particles creates the distinction between *proton* and *electron*: If the theory bearing on this particular kind of thick epistemic access is *right*, the distinction theoretically described is already there in the causal relations themselves that we use to gain thick epistemic access to something. The theory simply describes when those causal relations can be misleading and when they're not, and how to tell (if, i.e., the theory is specific enough to do this).

The reader may wonder if the distinction between a theory playing a constitutive role and its playing a descriptive role is a distinction with genuine content. Isn't every empirical theory (that's *true*) playing only a descriptive role? No, those that involve the application of mathematics may (at least partially) be playing what I'd describe as a constitutive role: The theory—at least as far as some of the mathematical posits are concerned—need not be squaring itself with what it seems to be talking about. I give examples of this in following chapters.

7

Posits and Existence

As I've said already, apart from the epistemic burdens posits have, it's an additional question what our ontological view of these things should be. Given the arguments of part I, it's a somewhat delicate matter to know about things that we take to exist. This is because things that we take to exist must not be made up out of whole cloth, and so they must satisfy the reliability and nontriviality conditions. Nicely enough, thick posits, items to which we have thick epistemic access, seem to easily navigate this particular epistemic hurdle. *Robustness* and *grounding* (as defined in items (1) and (4) in chapter 6) show that such objects can't be like fictional items or even like the contents of hallucinations—made up out of whole cloth—because (a) we can't stipulate the properties we attribute to things to which we have thick epistemic access, and (b) the properties of things to which we have thick epistemic access play a role in the explanations we give for how we have thick epistemic access to them. Fictional items violate both (a) and (b), and the contents of hallucinations, although they satisfy (a), violate (b). On the other hand, the epistemic role obligations that objects (which we can thickly access) have because of *grounding*, and the refinements and adjustments of thick epistemic access that *refinement* concerns, enable us to satisfy the reliability and nontriviality conditions with respect to thickly accessed objects. Thin posits, too, are items that satisfy (a) and (b) because the properties we attribute to them are theoretically bound to the thick posits that fall under the same predicates they fall under. Ultrathin posits, by contrast, seem nicely designed to be items that exist in no sense at all.

Quinean Rent and the Black Box Objection

As is well known, Quine took the payment of Quinean rent to suffice for an ontological commitment to a posit. One point to make right at the start is that if Quinean rent is to operate on its own as a decisive consideration about whether a posit exists or not, then it's an all-or-nothing affair. Every posit in a theory pays Quinean rent, or none of them do. So if the fact that the posits in an empirical theory pay Quinean rent implies that they exist, then it implies that they all exist, and this means that the mathematical posits included in that theory exist, as well. This, alone, is no objection to treating Quinean rent as licensing existence to the posits of a theory. But in point of fact, Quinean rent (alone) doesn't tell us that the posits of a theory, that is, the existential commitments of a theory, are the sorts of things we *should* take to exist (in any case). For it to do *this* requires there to be a criterion (for what a discourse commits us to) already in place. By the results of part I, such a criterion is that the posit in question falls under the "ontological independence" predicate, and so it must be determined whether Quinean rent (alone) provides such ontological independence.

For Quinean rent to provide ontological independence to those posits that pay it, it must enable them to satisfy the reliability and nontriviality conditions, and this it fails at. Satisfaction of the five Quinean virtues alone tells us nothing about how we know what we're taken to know about the posits (of a theory) that satisfy those virtues. All we know is that, *as a whole*, such a theory is, broadly speaking, *empirically adequate*, and this can happen even if *all* we can say about (all) the posits of such a theory (apart from what the theory tells us about them) is that they're the *relata* of the items in an empirically adequate theory. Empirical adequacy *does* tell us that, somehow or other, predictions of the theory are coordinated with something out there that is ontologically independent of us. But the theory provides no explanation for how the trick is turned; in particular, it provides no guidance for judging whether the theory manages this because *all* or only *some* (or none (!)) of its posits actually refer to things ontologically independent of us, and it provides no guidance for determining, if only some of the posits refer to items ontologically independent of us, which posits those are.

Precisely here I find myself breaking decisively with Quinean epistemology, because it's here that it despairs of the possibility of any further epistemological insights (and in this respect, *many* philosophers follow suit: Davidson, Putnam, Resnik, Rorty, etc.). This is because Quinean epistemology sees all ontological distinctions as *internal* to our web of beliefs: They're solely the result of grammatical distinctions in our language that are projected onto the world—a world that impinges on our theory only via the empirical adequacy of *whole* theories. In describing ontological distinctions as "internal" to our web of beliefs, what's meant isn't the mere tautological observation that, of course, *our* beliefs about ontology must be ones *we* hold, regardless of their subject matter. What's meant, more strongly (and *not* tautologically, although this latter thought is sometimes presented as if it's merely the tautology just mentioned—in order to enhance its digestibility[1]) is that we can't inspect sentences in our body of beliefs to distinguish those posits to which we're committed because of our theory-making needs from those posits to which we're committed, because they refer to items we take to be ontologically independent of us.

This is why the Boyd-Putnam "no miracle" argument for scientific realism isn't very forceful. A classical presentation of the argument can be found in Putnam (1975):

> The positive argument for realism is that it is the only philosophy that doesn't make the success of science a miracle. That terms in mature scientific theories typically refer (this formulation is due to Richard Boyd), that the theories accepted in a mature science are typically approximately true, that the same term can refer to the same thing even when it occurs in different theories— these statements are . . . part of the only scientific explanation of the success of science, and hence [are] part of any adequate description of science and its relations to its objects. (p. 73)

The problem can be easily pinpointed in the phrase "terms in mature scientific theories typically refer. . . . " We *may* agree that an instrumentalist take on an *entire* scientific theory leaves *inexplicable* the empirical success of that theory, but on the other hand, that same empirical success *cannot* be used to *indiscriminately* commit us to the referents (posits) of *every* term in that theory enjoying that success.

I need to stress that the argument against an inference to ontological commitments on the (mere) basis of the global virtues of a theory— which I'll call the *black box objection*—has very powerful ontological implications, so although I think the argument justified, I don't want to give the impression that I think it's trivially obvious.

As an illustration of its potential for providing us with ontological austerity (desert landscapes) in what we should be willing to commit ourselves to, imagine a case where the phenomena, plus the other usual theoretical considerations, allow only one empirically adequate theory. It might be thought that the *uniqueness* of such a theory shows the ontological independence of (all) the entities it existentially commits us to. After all, (by assumption) the data force us to only *one* theory (so, in particular, no stipulation is involved because there is no choice). But we can't, in this case, separate out the different factors causing the collapse onto only one theoretical option: Linguistic factors determining the form of notation the theory must adopt may be playing a role in what ontological commitments that theory has. Without thick epistemic access *to* the entities purported existentially quantified over, the theory (even a unique one) re-

---

[1] See, e.g., Putnam 1987, lecture 1, esp. p. 20.

mains an ontological "black box." We need to know that *the entities* are ontologically independent, and all that the uniqueness of a theory tells us is that we have no choice (for any number of reasons—including the possible failure to have developed certain mathematical breakthroughs) in what theory we adopt (and what that theory is quantifier committed to). But absence of choice in theory construction—which, after all, is still a global property of the whole theory—does not imply the ontological independence of the entities that the theory quantifies over. Inference to entities on the basis of a theory *without* thick (or thin) epistemic access to those entities is thus illegitimate.

Here's another application of the black box objection: Certain philosophers have stressed that a powerful factor in explanation is unification.[2] If a theory offers a pattern of argument that can be widely and successfully applied to explain (and predict) phenomena, and/or if the postulation of a "theoretical structure" (Friedman 1983, p. 236) with accompanying existential quantification leads to a substantial unification of the kinds of phenomena that can be described, then the theory has, in Friedman's words, "unifying power." And not only is this an intrinsically appealing property for theories to have, but if a theory U unifies various empirical applications of more specific theories $Sp_1, \ldots, Sp_n$ under itself, then it garners additional confirmation for itself over and above the various more specific theories; indeed, Sp_1, for example, garners more confirmation for *itself* by virtue of its deduction from U than it would all on its lonesome (because, via its attachment to U, it additionally gains the confirmation accrued to $Sp_2, \ldots, Sp_n$).

This much I won't argue with. Friedman (1983), however, draws a further ontological conclusion from this. Simplifying slightly,[3] the claim goes something like this: If, in order for U to unify $Sp_1, \ldots, Sp_n$, a kind of entity must be co-referred to (so that it appears in each $Sp_1, \ldots, Sp_n$—after possibly reconstruing some of the specific theories $Sp_1, \ldots, Sp_n$, and, of course, it also appears in U), then "we have no choice but to take it literally, to assign it a rightful place in the 'world' of physical reality" (p. 250).

Friedman contrasts how a posit's playing a role in unification in this sense should commit us to the physical reality of that posit versus the remaining posits in the theories in question, which are to be treated as merely part of the mathematical formalism accompanying $Sp_1, \ldots, Sp_n$, and U. But the reader should be suspicious of this move on three counts:

---

[2] Two prominent examples are Philip Kitcher (e.g., his 1981) and Michael Friedman (e.g., his 1983, chaps. 6 and 7, esp. pp. 236–50). Both philosophers acknowledge the recognition of this factor as significant for explanation on the part of earlier philosophers of science such as Hempel, Whewell, and others. See Kitcher 1981, and Friedman 1983, p. 242, n. 14. In what follows, I focus exclusively on Friedman's discussion.

[3] Friedman focuses on "structures"—such as absolute rest—not entities committed to. For our purposes, this doesn't matter: Such structures translate into entities committed to as well as (certain) truths about those entities.

First, the confirmation gained by theories this way applies *sententially*; that is, what's garnered, at best, are additional reasons to think the *theories* in question are true, and this, by itself, doesn't bear on the ontological commitments of such theories at all. Second, the kind of unification at work here seems widespread in *pure mathematics* itself—and is even prized there for its explanatory power. One feels one has a deeper understanding of various more specific phenomena, say, when one shows that what's really at work is the same group structure. Finally, and related to these first two points, I can raise the black box objection to this approach: Theoretical unification, no matter how valuable, does nothing to tell us that the virtues (confirmational and otherwise) of the resulting theory are due to some of its terms (ones essential to the unifying process) picking out what's physically real. Nothing Friedman says shows that certain terms being essential to the greater unifying power of the theories they appear in *forces* those terms to refer to real parts of the world.[4]

Ontological Commitments in Science

Physical theories, which are deeply mathematical in character, routinely impose global constraints on the universe that translate into "*existence*" conditions; this can allow pure theory to have quantifier commitments to other items in parts of the universe *inaccessible* to us. As I've shown in chapter 6, it's a subtle matter whether the theoretical links between what's implied to exist on the basis of theory and what we've forged thick epistemic access to is tight enough to justify the conclusion that what's theoretically posited is actually thin. Imagine, for example, that a particularly well-attested theory implies the existence of a certain subatomic particle, but that it also follows from that theory that the energies needed to actually forge thick epistemic access to such an object are (forever) vastly beyond our capacities. Physicists, as a community (so I claim), would *not* commit themselves to the existence of that particle on purely theoretical grounds—no matter how much empirical support in other respects such a theory had.

The reason for this, and why it contrasts so starkly with the virus case, is that the theories about subatomic particles are on the ground floor (scientifically speaking). This means that apart from the theory of subatomic particles itself, there's little theory to bring to bear on the question of whether we should take such a subatomic particle to exist. What makes the other examples I've mentioned different is both that there is empirical evidence—the constant discovery of new kinds of, say,

---

[4] Friedman does seem to be helping himself to a refined version of an indispensability argument: Terms *essential* to how U is linked to the theories it unifies are the ones we're to take as picking out real pieces of the world. But, really, this doesn't do anything more to show that such terms refer to what's ontologically independent than do considerations based on the ordinary indispensability argument.

viruses (or chemicals)—and that there is a body of underlying theory about viruses and chemicals (and how variable they can be) that gives us good reason to think that such different kinds are out there. This is hardly the same as a case of a single inaccessible type of subatomic particle.[5]

Theologians, or some of them, anyway, are perhaps different in this respect. Proofs of God, if any could be found, *would* amount to the acceptance of a rather unique posit on the basis of purely theoretical virtues (the virtues, in particular, of an applied logic that the more religious among us would otherwise be loathe to change). But theologians, after all, are a desperate lot.

Certain philosophers, however, seem as willing as theologians to employ methods of pure logic to satisfy their ontological aspirations. As I mentioned in chapter 3, in the section titled "Some Important Illustrations," it's widespread in contemporary metaphysics to extract commitments to various types of object on the basis of the logical form of certain sentences. Often, the resulting commitments are taken to be abstracta, and I've already noted my objections to finding oneself so committed to such items, at least on a basis like this. But these commitments are also sometimes seen as concreta, and here one can only say that scientific methodology doesn't seem to allow a commitment to such entities, without thick epistemic access to other items that fall under the same kind-term, and unless the theory of this kind-term gives us good reasons to think it has instances to which we *don't* have thick epistemic access.[6]

A natural question that arises at this juncture is what status the mathematical posits that arise in applied mathematics have. I'll argue that such

---

[5] Of course, if there were numerous subatomic particles—far more numerous, I mean, than, say, the number of hadrons—*and* we regularly discovered new ones on the basis of our theory, *and* our theory was one that was extremely well confirmed, the epistemic situation would change. Consider, however, the following from Brown et al. 1997:

> A major goal of the PETRA storage ring, and of a similar, competing collider at SLAC named PEP . . . was to discover the top quark required by the emerging Standard Model. At the time, phenomenological models suggested that the top quark's mass might fall in the 10–20 GeV range accessible at these machines, but early experiments failed to find any evidence for its existence. So firm was the belief in the Standard Model by 1980, however, that this failure did not lead to serious doubt about the theory's validity. Physicists widely supposed that the top quark had to be more massive than could be produced at PETRA and PEP, and that it would eventually be found at higher energies. (p. 23)

Notice that it isn't that the existence of the top quark is, in principle, out of the range of empirical verification: It's that it's out of the range of specific colliders—and in drawing this conclusion, physicists are contravening only phenomenological models.

[6] Mental objects and events, on some views of these (as chapter 3, "Some Important Illustrations," indicated) are posited solely via logical form; so are fictional objects (on some views of *these*) and Lewis's possible worlds. Regarding the last, it's worth noting that *our* thick epistemic access to, as it were, *our* world, hardly suffices to satisfy the above requirements as far as *other* possible worlds are concerned. In any case, logical-form considerations won't do for such ambitious ontological purposes, as I argued in chapter 3.

posits fall neatly into two categories: Either they're a kind of item that fails to have close specific theoretical ties to items we have thick epistemic access to, and then they're simply ultrathin posits, or they're a kind of item that satisfies the constraints on thick and thin posits, and then they—the mathematical posits themselves—are thick or thin (or both). The interesting question, to which I'll turn in chapter 8, is which ones are which, and how we tell.[7]

Ontological Identifications

From now on, I identify the following three types of posit: (i) ontologically independent posits, (ii) causally efficacious posits,[8] and (iii) posits that exist. Any of these notions could, in principle at least, diverge in its extension* from those of the others, but I'll argue that they all coincide.

I'll also identify, in what follows, (i′) posits that are ontologically dependent on linguistic or psychological processes, (ii′) posits that don't exist, (iii′) ultrathin posits,[9] and (iv′) causally inert posits.

How are the first set of identifications motivated? What we already have in place is the identification of ontological independence (in *our* linguistic community, anyway) as the criterion for what exists. This, in turn, leads to a characterization of how we identify what we take to be ontologically independent of us, that is, the reliability and nontriviality constraints of chapter 3. An excursion into (naturalized) epistemology then reveals that the only way *we* have of learning about a kind of thing ontologically independent of us is to forge thick epistemic access to some of its instances. In turn, and this engages us in metaphysical characterizations of what we can take to be ontologically independent of us, it's

---

[7] I'm making a claim here that will be further substantiated during the course of part II. After all, it may seem to some that space-time points, say, are neither thick nor thin in my (current) senses of these terms, nor ultrathin. Later I'll undercut the cogency of this option.

[8] The identification of type (ii) with types (i) and (iii) commits me to what has come to be called the *Eleatic Principle*. See Oddie 1982, where this term was coined. Also see Armstrong 1978 and Colyvan 2001.

[9] A terminological caveat intrudes here: Consider the term "chimera"—as it is taken to refer* to the mythological creaure—and imagine that we (or the ancient Greeks, anyway) once believed in these things. In saying (as we now do) that "there are *no* chimeras—in this sense," we're treating "chimera" as a thick or thin posit that has (unequivocally) failed to live up to its epistemic obligations. We can also treat the term as ultrathin. Then we say, "Chimeras are mythological beasts believed in by ancient Greeks." So the distinction between ultrathin posits and thick (or thin) posits that have failed at satisfying their epistemic burdens, is a *soft* one. This is no surprise because nothing at all rides on drawing sharp distinctions in this way between kinds of posits we take not to exist in any sense at all. The status of a term like "chimera" in usage turns on pragmatic issues—whether we're informing a person about truths that involve the term "chimera" or whether we're trying to prevent a misguided view about chimeras—as so understood—that takes them to be *real*. I'll presume that any term, purporting to be thick or thin, that unequivocally fails at fulfilling its epistemic burdens is reclassified as ultrathin—but as I've said, nothing much turns on this.

determined that such items are causally efficacious—for it's only by developing causal connections *of the right sort* to things that we can forge thick epistemic access to them.[10]

It seems clear that the foregoing considerations aren't strong enough to provide the *metaphysical* identifications (i) – (iii), because although ontological independence and existence should be identified if the considerations in part I are correct, the identification of these with causal efficacy needs additional argument: It seems it's in principle possible for there to be something ontologically independent (out there) and yet causally idle, that is, without causal powers of *any* sort.

On the other hand, some philosophers may think that in adopting the Eleatic Principle, the claim that everything that exists has causal powers, one is engaged in a kind of induction that's commonplace in ordinary science: Scientists are willing to presume—on the basis of quite limited data—that such and such theoretical constraints exhaust the possibilities of what there is. Scientists are quite sure, for example, that nothing can move faster than the speed of light; this kind of claim, of course, is always subject to the condition that our theories are right (one can be wrong about anything), but that condition aside, there's no reason why metaphysical claims about causal efficacy should be treated differently from other claims in science.

A way of resolving this is to consider more closely how certain sorts of ontological conditions *are* ratified in science: Until now, I've been almost entirely concerned with how ontological commitments in science are established; I've argued, for example, that thick epistemic access to instances of a kind of item are needed before we're willing to claim that items of that kind exist.[11] But there is another aspect to ontology that bears on the issue at hand, what I'll call *ontological closure conditions*. The periodic table in chemistry is an excellent example of such: Not only does it supply a list, and rules of thumb, governing what elements there can be, and what properties they're likely to have, but it also *excludes*, quite explicitly, the possibility of other sorts of elements. It's easy to see how it does this: The atoms of such elements are individuated by the number of protons in their nuclei; because protons are discrete items, this puts severe constraints on what elements are *possible*.[12]

Something similar happens with subatomic particles: A sort of "periodic table" is also established of what sorts of properties can vary among the fundamental constituents of these things (charge, spin, flavor, etc.),

---

[10] Although we don't, in practice, forge thick epistemic access to *every* instance of a kind, we're still justified, on the basis of the theory about such a kind-term, in inferring that the remaining instances (the thin posits falling under that kind) are also causally efficacious.

[11] This claim, of course, is subject to the considerations about subkinds and thin posits discussed in chapter 6.

[12] In particular, there is, *at most*, one element corresponding to each positive integer. I'm not, of course, speaking of muonic or exotic atoms.

and this puts constraints on how many kinds of subatomic particles there can be.[13]

Scientific *theory* is clearly used to supply these closure conditions: Only such things can exist. Why "can" (and not "do")? Because theory often can't go on to tell us that therefore such things *do* exist: As I've argued already, what's additionally needed is thick epistemic access to (some of) the items of the kind in question before we're convinced of that.[14] And it's easy to see why theory, in general, should not be allowed to provide ontological credentials without the involvement of thick epistemic access. Apart from the fact that any theory could be wrong,[15] theories don't invariably tell us that what the theory allows to exist must exist simply because other constraints may come in that exclude the possibility the theory seems to allow.

Consider again the case of the periodic table: That table is simply a (periodic) ordering of possible elements on the basis of the number of protons in a nucleus and the corresponding number of electrons in the shells surrounding that nucleus. But perhaps, if we worked out the fundamental force laws governing the contents of such nuclei, *and* discovered tractable mathematics that enabled us to deduce from those laws what nuclei are possible, we could have discovered that certain numbers of protons in nuclei are *ruled out*. That is, specific theories can often be dictated to by broader and more fundamental theories that rule out possibilities the more specific theory allows.[16]

---

[13] Brown et al. 1997 write: "The model is referred to as 'standard', because it provides a theory of fundamental constituents—an ontological basis for describing the structure and behavior of all forms of matter (gravitation excepted), including atoms, nuclei, strange particles, and so on" (p. 3). That this "periodic table" doesn't include gravitational interactions is a main motivation for the exploration of string theory. Indeed, attention is centered on closure conditions, such as this periodic table of fundamental particles, when theories change. The introduction of string theory *allows* additional sorts of objects to which (in principle) thick epistemic access can be forged (although it still limits what's possible). One way, thus, to provide evidence for string theory over its more restrictive competitors is to forge thick epistemic access to these predicted items.

[14] We find in the table of contents of the September 1998 issue of *Scientific American* (p. 5) under Making New Elements: "If they exist, elements 114 and beyond may prove surprisingly stable."

[15] Does the fact that a theory can be wrong contribute to why thick epistemic access plays a more fundamental epistemic role than do theoretical considerations in what we take to exist? If so, why should it (since we can be wrong about what we think we have thick epistemic access to, just as our theories can be wrong)? The reasons for this turn on conditions (1) – (4) of thick epistemic access (defined in chapter 6), which make it epistemically more reliable, in general, than scientific theory.

[16] A neat example of this is supplied by evolutionary biology and chemistry. That, e.g., carbon-based life is possible, and that lithium-based life is not, is dictated to biology by chemical theory. That other metal compounds can play a similar role in blood that hemoglobin plays in the blood of mammals is also dictated by chemistry, and it also tells us that these other compounds won't be as efficient, and that therefore we're unlikely to find them in anything other than invertebrates.

In a case where the scientific theory is on the "ground floor," as it were, so that principles from below—from a more fundamental science—can't be brought to bear on the theory, we may still be unable to say for sure what it allows to exist and what it rules out simply because the theory may be too intractable to derive the needed consequences from. So we're once again, even if the theory is one we're quite committed to, forced to thick epistemic access to verify what there is.

The methodological fear one has about ontological closure conditions in any science is straightforward and familiar: It's that it always seems empirically *possible* for something to come along that the theory has apparently ruled out. Our access to what's out there is so limited, so parochial, and consequently, our evidence for our theories is so limited and parochial, that we're always in danger of having our theories overthrown, not only because what they take to exist *doesn't*, but also because what they take not to exist *does*.

This is, however, nothing other than the problem of induction, and there is, as Hume knew, nothing to be done about it when it arises in this global form: We can be wrong about *anything*. Nevertheless, scientists can design their theories, and test them so that if the theory, having passed such tests, still subsequently proves to be wrong, it's because an epistemic catastrophe has occurred (something unexpected has arisen—perhaps experimentally—that we had no way of testing for or anticipating before we ran across it) versus the violation of the theory simply being something that was easily foreseen and (possibly) overlooked.[17]

---

[17] The same principle is at work in building a house well. One cannot design houses to withstand *every* possible disaster—but better-built houses withstand ordinary adverse events better than do ill-built ones.

Weinberg (1993), when considering "the final theory" in physics—the theory that will explain everything—worries about the possibility that the question "But why is *this* theory *true?*" will still be with us. He hopes to make the final theory more satisfying than a mere brute fact by showing

> that the final theory, though not logically inevitable, is logically *isolated*. That is . . . although we shall always be able to imagine other theories that are totally different from the true final theory . . . the final theory . . . is so rigid that there is no way to modify it by a small amount without the theory leading to logical absurdities. In a logically isolated theory every constant of nature could be calculated from first principles; a small change in the value of any constant would destroy the consistency of the theory. (p. 236)

It's hard to see why a logically isolated theory (in this sense) is more satisfying as far as explaining why things are like *this* (rather than like *that*) is concerned. All a logically isolated theory explains is why, *given* that the theory *is* true, *we* couldn't have come up with *something else* (given the data and the mathematical tools we luckily had access to). But a logically isolated theory does have genuine *epistemic* advantages: Even with small differences in the data, we're still forced to the same theory. Only an epistemic catastrophe, something that shows us that we're radically mistaken in a fundamental way that implies, among other things, that we can't even preserve most of the evidence we've gathered in support of that theory (or of its successors), would overthrow a logically isolated theory.

"Epistemic catastrophe" is perhaps melodramatic language. It's a truism that our ability to test the numerical implications of our theories improves incrementally, as we enhance and develop instruments. Such refined tests can always (we recognize) overthrow a theory simply because it fails to hold under refined experimental conditions.

I hope it's clear that I've no intention of being dismissive of skepticism, at least skepticism about induction. Despite the dangers of induction, however, it seems clear that science is willing to take risks: Given that we have very good reasons to take a theory to be true, we also take ourselves to have very good reasons to foreclose on what exists solely on the basis of that theory.

With this sketch of how ontological closure conditions are established in science, let's see how they can be used to motivate the identification of causal efficacy with ontological independence (and existence). The first move is to accept the principle that, in Quine's words, the job of (fundamental) physics is full coverage. This isn't, as I understand it, the acceptance of the typical forms of reductionism of the special sciences to physics that (many) philosophers have been committed to: It's neither a commitment to strong reductionism, the claim that the kind-terms of the special sciences can be defined in physical terms and that the laws of the special sciences can be deduced from the laws of physics; nor is it a commitment to weaker versions of reductionism, which are typically couched in terms of one or another kind of *supervenience*, a view, for example, that instances of a (special science) kind are required to correspond to (in some way) or fall under (one or another) kind-term of fundamental physics.[18]

What *is* required, so I claim, in describing physics as having the job of "full coverage," is (1) that physical law is to be understood as applying (in principle) to everything. In practice, of course, the specific application of physical laws to the macro-objects that the special sciences are typically about is generally ruled out by intractability considerations. But in those special cases where such laws *can* be applied successfully, there should be no obstructions arising from the laws and truths of the special sciences in so doing. (If a wingless rodent falls out of an airplane, we'll treat it as a falling body, and not worry about the possibility of emergent aerodynamic phenomena because it's a *rodent*.) This condition I take not only to hold of (fundamental) physics with respect to (all) the other sciences, but to hold of *any* two sciences, where the domain of application of one includes the domain of application of the other, and where, furthermore, the first is more fundamental.[19]

---

[18] The reductionist and supervenience literature is gigantic. For details on my views about the various kinds of reductionism, why I reject them, and some references, see my 2000b, pt. I, § 3. Also see n. 20. For an excellent presentation of the issues surrounding supervenience, see Kim 1993.

[19] Some versions of physicalism involve the view that what there is is in some sense to be supplied (solely) by physics. I have rashly and in passing word processed remarks supportive of such a view (e.g., in my 2000b, pp. 45–6). I say a little more about my current views on this in chapter 9.

My second claim is more radical and is one that I won't fully argue for here: (2) that the previous condition is the *only* condition that can be placed on the sciences with respect to one another, and that nevertheless, such a condition suffices for a commitment to "physicalism," the view that in a perfectly respectable sense, physics, in principle, gives us (all) the laws for everything there is.[20]

For our current purposes, the second claim isn't needed because (1) any physicalist is at least committed to my first claim (and a number of nonphysicalist positions are committed to it, as well), and (2) in any case, my first claim is strong enough to support the metaphysical identification of causal efficacy with ontological independence and existence that I want to argue for.[21]

Here's how to do it (the quick and dirty way): Ontological closure conditions from physics tell us that physical laws apply (in principle) to everything there is. But anything causally inert won't be amenable to physical law; therefore, such things are ruled out.

Caveats: First, some may be worried that quantum mechanics, Bell's theorem in particular, directly undermines causality. It doesn't—what it

---

[20] Some argument for this position may be found in my 2000b, pt. I, § 3. Cosmides et al. (1992) offer the following consistency principle on the sciences, which they call "conceptual integration": "[T]he various disciplines within the behavioral and social sciences should make themselves mutually consistent, and consistent with what is known in the natural sciences as well.... A conceptually integrated theory is one framed so that it is compatible with data and theory from other relevant fields" (p. 4). They claim that the natural sciences already obey this principle, but they fault evolutionary biology, psychology, psychiatry, anthropology, sociology, history, and economics for not doing so. They write: "As a result, one finds evolutionary biologists positing cognitive processes that could not possibly solve the adaptive problem under consideration, psychologists proposing psychological mechanisms that could never have evolved, and anthropologists making implicit assumptions about the human mind that are known to be false (p. 4)." This principle strikes me as equivalent to the one I've just offered, except for the following caveat that I insist upon: It's perfectly acceptable for a special science to be (entirely) constituted of laws that are "idealized" in such a way that (1) we recognize these laws (one and all) to be strictly false (from the vantage point of the truths and laws of a more fundamental science), and yet (2) we find the special science, as formulated, indispensable because specific application of the broader principles (of the more fundamental science) to the specific domain of the special science is intractable. Also, it is allowable for the laws of a special science to be quantifier-committed to things that, strictly speaking, don't exist. Arguably, linguistics (or a great deal of it, anyway) is such a special science on both counts.

This caveat means, as I've indicated, that although one mustn't allow such a special science to impede the implementation of a more fundamental science to the former's domain of application, one shouldn't fault a special science *merely* for positing processes that don't exist or for invoking principles "known to be false." So too, and for the same reasons, one shouldn't fault a special science *merely* for positing entities that fail to be definable in terms of or supervenient upon the entities (of an underlying science) that we take to be the only ones that "really exist."

[21] Many philosophers try to establish the Eleatic Principle not via a consideration of the methodology of science, as I'm doing now, but by the use of purer philosophical tools such as general considerations about the structure of inference to the best explanation or Occam's Razor. I've already (in chapter 4) given reasons for thinking this route won't work.

does undermine, if anything, is some such principle as "Every event has a cause" or, less vaguely, "every correlation has a common cause." But it's hard to see why the collapse of *that* principle affects the metaphysical status of causation.[22]

Second, one may wonder why a similar sort of argument can't be used to establish the co-extensiveness* of the disjunction of thickness and thinness with what exists. After all, don't we take all the posits of physics to be either thick or thin? If so, we can piggy-back on the ontological closure condition that tells us that everything physics says there is is everything there is to the conclusion that (therefore) everything is either a thick or a thin posit.

I can't fault this argument on the grounds that it takes current physics to be true: No ontological closure condition survives the consideration that our theories might be false—that's a given. The problem is with the next two steps. We do *take* all the posits of physics *that we specifically know to exist* to be thick or thin, but that's not enough for us to claim that all the posits of physics *are* thick or thin. Suppose I stubbornly refuse to accept the existence of anything I can't see. Even so, I can't claim, from that, that everything that exists is something that I *can* see. I can't claim *that* unless I also believe that anything I *can't* see *doesn't* exist. Only then is "seeing" (factively understood) operating as a criterion (a necessary *and* sufficient condition) for what exists. But if I merely claim not to know about the existence of anything unless I see it, then seeing is only a sufficient condition for what exists, not a necessary one.

The identifications of causal efficacy with ontological independence and existence, and in particular, how the Eleatic Principle is justified, haven't turned on epistemic arguments involving causality; I've *not* argued, for example, that if there were causally idle entities, we'd have no reason to believe in them. Nor have I taken the approach that Colyvan (2001, pp. 41–2) calls the Inductive Argument: All the entities we're currently committed to the existence of are causally efficacious, and therefore we've good (inductive) reasons to think all entities are causally efficacious. *That* argument, as I've just shown, fails to recognize that induction can't operate solely by listing things that we take to exist and looking for common properties. This only gives us, at best, sufficient conditions. Ontological closure conditions are needed to give necessary conditions.

What sort of ontological closure condition would be needed to establish the co-extensiveness* of the disjunction of thinness and thickness

---

[22] Cartwright (1989, chap. 6) argues that it doesn't undercut the common cause principle: Other assumptions needed to draw this conclusion go first. The issue is delicate enough and complicated enough to require much more than I can do now, as the very large literature on this topic indicates. So I'm only staking out my position on this (without argument).

Another way that quantum mechanics may seem to undermine causality is that measurement, on the Copenhagen view, for example, requires the "collapse of the wave function." What this means is that an otherwise completely deterministic process (when "mea-

with what exists? Something to the effect that nothing exists that falls outside the nomenclature of our theories. Motivating such an ontological closure condition is quite difficult. Suppose (contrary to fact) that we had an ontological closure condition for the kinds of items admissible in the language of physics.[23] Then we would still require (a) that the posits that we take to fall under the nomenclature are either thick or thin, and (b) that by virtue of (a), for every other sort of posit p (of the special sciences), either (b1) we have thick epistemic access to p, or (b2) we have enough of a theoretical grip on p so that it's a thin posit.

This strikes me as implausible in the extreme. For example, thick epistemic access to subatomic particles plus access to the theory of such things doesn't *by itself* suffice either for thick epistemic access to the lizards composed of those subatomic particles or even for access to a biological theory of lizards, and this is for reasons that I've already discussed: (1) Thick epistemic access to something doesn't suffice—usually—to supply thick epistemic access to objects that thing is a part of (or *vice versa*), and (2) the intractability of the theory of subatomic particles prevents the deduction of theories of macro-objects.

Colyvan (2001), I should say, offers his own objection to the inductive strategy for establishing the Eleatic Principle:

> [T]here are many other properties that the uncontroversially real entities share and that the uncontroversially unreal entities lack. All the uncontroversially real entities are spatio-temporally located, for instance.... The reality of space-time points hangs crucially on whether it's causal efficacy or spatio-temporal location that is the important property. Similarly, we could opt for the property of having a positive rest mass as the mark of the real and again the demarcation would be different. (pp. 41–2)

This gives a false picture of the options available for the common properties held by everything we take to exist. In particular, it seems quite odd to take, for example, positive rest mass to be a criterion for what exists. *Light* is hardly like the contents of hallucinations, or like numbers, for that matter, and so it's hard to see why it should be assigned a status on

---

sured") randomly jumps to a new state in a way that's not characterized by the dynamical laws (Schrödinger's equation) governing that wave function. I think this can be shown, contrary to appearances, not to undermine causality, either, but my reasons for thinking this also take us too far afield.

[23] I did *not* presuppose such a thing, note, to establish the Eleatic Principle. Such a closure condition would require a limited nomenclature in physics from which all the other terminology of physics would be *definable*. I doubt this would be possible even if, say, we discover the "theory of everything." In brief, this is because the applications of physics to what exists allows that application to be to items characterized in ways that fall outside physics—such will always be necessary because of the introduction of idealizations to make the application of physics tractable—and such idealizations can (and usually do) involve posits that needn't be strictly part of the "theory of everything." See, for numerous examples of this, Truesdell 1991.

a par with such things.[24] Space-time points are another matter: If they're the posits of one or another branch of pure mathematics, then, of course, they're like fictional objects, as I've already argued. But if we're speaking of the posits of an empirical theory, then it remains to be determined whether they're ontologically independent or not. Surely the simple assignment of a location for them won't confer ontological independence to them all by itself.[25]

The foregoing considerations about ontological closure conditions, although of interest in their own right, have been brought on stage at this point in order to defend the identification of causal efficacy with existence, specifically, to rule out the possibility of causally inert objects that nevertheless exist. But there is a second way that the identification of causal efficacy with existence can be troubled: if Kant's challenge[26] is brought against it. I'm not concerned with the attribution of causal efficacy to fictional entities *in fiction*: As I mentioned when I first raised Kant's challenge, attributions to fictional entities, at least when they occur within fiction, needn't be taken to be true. The same, however, isn't true if such attributions are made to nonexistent entities within the context of *applied mathematics*, because in that case, the attributions, as I showed in chapters 1 and 2, must be taken to be *true*. I'm anticipating an analysis that I carry out later in this book, but it's my contention that point-masses and space-time, for example, don't exist. Nevertheless, it may seem that we attribute causal powers to such things—that, for example, space-time (in the context of general relativity) itself has energy, and that point-masses, in the context of Newtonian physics, exert gravitational force upon each other.

My response to this may look like a proverbial biting of the bullet (but, I think, it really isn't). I first claim that attributions of causation are factive: When we describe something as the cause of something else, we commit ourselves to the existence of the first something (and the second something, for that matter). This is, perhaps, not particularly controversial, and so I won't dwell on it. My second claim should also not appear particularly controversial—but it will lead to something in need of further argument pretty quickly: This is that there aren't any idioms in *physics* that, all by themselves, anyway, betray the presence of causation. Rather, causation is something *imposed* upon that formalism when we apply it. In saying this, I'm *not* making the Humean-sounding claim that causation is a projection onto the world—or onto the scientific description of that world—and isn't therefore in that world. Rather, I treat causation much

---

[24] Recent experiments have been carried out to determine whether (certain types of) neutrinos have positive rest mass. It's odd to learn that it's "uncontroversial" that what was actually being determined is whether those types of neutrinos exist or not.

[25] The contents of hallucinations, notoriously, are often unpleasantly located in space and time (e.g., the malevolent trolls *standing right in front of you* . . . while, that is, you're under the effects of certain drugs), but this doesn't confer existence on *them* either.

[26] See chapter 4.

the way I treat *ontological commitment*—it's something that we don't read off of scientific doctrine in any straightforward way—but this doesn't mean that we deny its presence (any more, i.e., than we deny the existence of *real objects* just because recognizing what aspects of scientific doctrine refer to real objects isn't straightforwardly read off of scientific doctrine either).

Unfortunately, these points about causation are deserving of extensive elaboration, and I can't undertake that in *this* book. Two last points have to suffice for now. The first is that we *don't* take something as causally efficacious unless we take it to exist (and so we don't take point-masses or space-time as causally efficacious but, at best, as *mathematical proxies* for something else that *is* causally efficacious), and the second is that we *shouldn't* treat "because" as a linguistic symptom of the presence of the attribution of causation, as the widespread presence of that idiom for purely logical purposes, and in mathematics, already shows.

In closing, I'd like to discuss several other identifications among posits that we make. First, the justifications for the identifications among (i′) – (iv′) are straightforward—given, anyway, the arguments of part I—because in their case we're concerned not with items that might have existed (but turned out not to) but only with the posits of our own theories that don't exist. But the reader may be curious how thickness and thinness operate within these classifications, if at all. There is something interesting involved here, and so I'll take a page or so to discuss them.

Thick and thin posits are to be identified with those posits in our body of beliefs that we *take* to exist, with one twisted qualification: We also believe that there are items (items that exist, that is) that are neither thick nor thin. So if predicates characterizing thickness and thinness (say, $\mathbf{Th}_c$ and $\mathbf{Th}_n$, respectively) are part of the language of our body of beliefs (and why shouldn't they be?), then for any predicate of our language P (excluding the existence, thinness, and thickness predicates and those taken to be co-extensive* with it), and where $\mathbf{E}$ is the existence predicate, we take the following claims to be true:

$(x)[(Px\ \&\ \mathbf{E}x) \rightarrow (\mathbf{Th}_c x \vee \mathbf{Th}_n x)]$, and
$(\exists x)\ (\neg\ (\mathbf{Th}_c x\ \&\ \mathbf{Th}_n x)\ \&\ \mathbf{E}x)$

As I've mentioned, the Quinean criterion applies in the same way to our body of beliefs and to other ones, provided they're based on first-order logic.[27] But this isn't quite the case here, both with alien bodies of

---

[27] Well, it does anyway, on my construal of how it operates (see chapter 3). There is a close analogy here: To evaluate what others are committed to, we must focus not on what their existential quantifier actually picks out but on what they take it to pick out (what existential sentences they're committed to—not which existential sentences are *true*). Similarly, when evaluating ontological commitments via an existence predicate, we must focus on what they take that existence predicate to pick out, not what it *actually* picks out. Quine's approach and the existence predicate approach look more disanalogous than they

beliefs and with our own earlier bodies of beliefs (which we've subsequently deserted), because thick access, as I've indicated, in our own case is factive (and notice that the terminological choice to make thick epistemic access factive doesn't cause problems with our own body of beliefs because we take ourselves to have really observed what we think we've observed). But things come apart when we describe others as having been mistaken about what they'd thought they'd observed, because if someone *thinks* she's seen elves, then that someone is committed to elves—regardless of whether she's *actually* seen elves.

Notice how this differs from how we evaluate inferential mistakes: If someone makes an (invalid) inference to the existence of *B*s from a body of doctrine, we're prone to say that such a person only *thinks* he's committed to *B*s—not that he's really committed to *B*s. But observation, as I've just noted, is different: If someone only thinks he's seen elves (but really hasn't) then he *is* committed to elves.[28] So it's the nonfactive use of observation (and thick epistemic access) that's relevant to determining what those committed to an alien body of beliefs are committed to believing exists—what they *take* themselves to have observed (and instrumentally interacted with), not what they've actually observed (and instrumentally interacted with).[29]

Why not then adopt the nonfactive use of "observe" and "thick epistemic access" and thus allow ontological commitment to be captured in both alien and domestic bodies of beliefs by instantiations by the same predicates ("thick" and "thin")? The answer is because it's important for our kind terms—except when they're in a restricted class of terms that pick out artificial items of a certain sort—to be "criterion transcendent," to be terms that we can be mistaken about the extensions of.[30]

---

actually are only because where a kind-term is involved, one is apt to interpret it according to how it has been translated (by our lights) rather than by what native speakers take it to pick out.

[28] Here is yet another example of how ordinary intuition treats observation as epistemically more fundamental than inference.

[29] This difference between first and third person with respect to the evaluation of ontological commitments corresponds to a similar difference between first and third person observation-attributions. Someone can say, when describing the effects of a drug: "The drug is working: I'm really seeing elves," even if her judgment is unaffected and she *knows* there are no elves in front of her (*and* that she's hallucinating). It's much less comfortable to say this in the third person: "She's really *seeing* elves," even with the stress where I've placed it, is too close to ontologically underwriting the claim that there *are* elves. Notice we can't say that "she *thinks* she's seeing elves," either—that's only appropriate if she thinks she might be seeing *actual* elves. We have to say, "She's hallucinating elves" or something like that.

[30] See my 2000b, pt. IV, § 3, for details.

8

Applying Mathematics: Two Models

I turn now, and pretty much for the rest of the book, to a study of a small number of restricted (but informative) examples of applied mathematics. But first we should ask if a general characterization of how mathematics is applied to an empirical domain can be given. I think one can, and measurement theory is a nice initial guide.

I first present (a slightly modified) version of a description from measurement theory of how mathematics is applied, give some examples, and describe some aspects of this approach. I then offer a different characterization of how mathematics is applied, one that can handle the other sorts of cases of applications of mathematics that I want to talk about.[1]

To this end, let's start with an algorithmic system S (Peano arithmetic, hereafter PA, is a good example),[2] and let's also assume we have a nonmathematical characterization E of some empirical domain, where this (in practice, not axiomatized) characterization is one of a collection of objects, as well as relations and properties holding between and of such objects.

Taking a tip from Krantz et al. (1971), a sufficient condition for the applicability of S to E is this: There is a (multiply valued) mapping, g, which in general isn't an "onto" mapping, and which (1) maps the con-

---

[1] Some of the material to follow is drawn and modified from my 2000a, pp. 216–20.

[2] The reader, however, can use any axiom system as an example of an "algorithmic system." Axiom systems are language based, but I understand algorithmic systems to be broader in scope: *Any* system of notation, language-based, diagrammatic, etc., with a decidable proof-recognition procedure is an algorithmic system. See my 1994, pt. II, where algorithmic systems are described, and my forthcoming(a), where Euclidean geometry is characterized as an algorithmic system that is (at least partially) not language based.

160

stant terms of S to constant terms of E, (2) maps the function terms of S to function terms of E, and (3) maps the n-place predicates of S to n-place predicates of E.

> *Definition*: g is *empirically adequate* if g induces on the sentences of S a mapping from the theorems of S to (not necessarily all) the truths of E.

Here are two initial examples: (a) Let E be the domain of line drawings on a very large piece of paper supported by a (relatively) flat surface. Let S be Euclidean geometry. (b) Let E be a set of straight rigid rods susceptible both to a "concatanation" relation (a placing of rods end to end) and to a "less than or equal" relation (understood to apply to rods with respect to their lengths). Let S be the language of addition and ordering on whole numbers.

Elucidations

1. The reader familiar with Krantz et al. (1971) may be more struck by the differences between their approach and mine than with the supposed similarities. Measurement theory officially takes homomorphisms of empirical domains into (intended) models of mathematical systems as its subject matter. But this is less of a difference than it appears. A characterization of an empirical domain often involves a rich set of properties and relations taken to hold of objects in that domain, some of which are ignored when a particular branch of mathematics is applied to it. (E.g., when considering the measurements and sums of the lengths of rigid rods, we ignore their colors.) This practice, of course, can be finessed in standard measurement theory by considering only a subset of the properties and relations taken to hold of items in an empirical domain. On the other hand, the application of a mathematical subject matter to an empirical domain often involves treating distinct items in a domain in mathematically similar ways (e.g., two rigid rods can be taken to have the same length). By mapping the empirical domain to the mathematical domain, the standard approach need only concern itself with single-valued functions that aren't one-to-one rather than with multiply valued functions that aren't mapped onto the range, as with my approach. There isn't much of a difference, mathematically speaking.

A second apparent difference between standard measurement theory and my approach is that measurement theory is officially an application of *model theory*; that is, it's not mappings of languages used to characterize empirical domains to languages of (applied) algorithmic *systems* that are studied in measurement theory, but mappings of the items (and properties and relations of such items) to objects (and properties and relations of such objects) in intended interpretations of algorithmic systems. For example, instead of studying, say, mappings of the language describing rigid rods to the language of number theory, what are studied are mappings of rods and their lengths to numbers and relations among them, for example, the addition relation.

I've no objection (in principle, anyway) to this model-theoretic approach, provided it doesn't give the (wrong) impression that using models to characterize how mathematics is applied empirically presupposes the existence of mathematical abstracta. After all, it's not the language and theorems of mathematics that are described in applications (on this approach); it's the mathematical objects themselves (plus their relations and properties) that are so described. I think the model-theoretic approach *doesn't* presuppose mathematical objects in any case although it certainly presupposes a (meta-)language that's *quantifier* committed to such things. As I showed in part I, pending a criterion for what a discourse is ontologically committed to that weds ontological commitment to existential commitment, this won't threaten a commitment to abstracta. Despite this surety, it's perhaps informative that an analysis of the application of mathematics can be executed solely in terms of algorithmic systems and not via purported mathematical objects.

2. E's characterization of an empirical domain need not be one we take to be "right," or "true," or even particularly *accurate* as a description of the ontology of the phenomena to which S is applied. To illustrate, consider an application of Euclidean geometry to ink figures on a flat piece of paper. What's the ontology of the empirical domain here? Not implausibly (given current scientific views), we might take it to be one of subatomic fields in a relativistic space-time. Leaving aside the sticky question of exactly how the implicit unification of the quantum-theoretical domain and that of relativistic space-time physics is to be achieved, we obviously don't recognize the applicability of Euclidean geometry in this case either by way of a mapping of this poorly understood physicalist ontology to that of Euclidean geometry (as on the first approach) *or* by way of a mapping of the language of Euclidean geometry into that of the language of fundamental physics (as on the second approach). For one thing, we can apply Euclidean mathematics to figures on a piece of paper without knowing anything more than that the empirical domain contains *figures* on a surface (piece of paper). One way or another, we've done exactly this for over 2,000 years—if not on paper, then on papyrus, or in sand, or *whatever*. Apart from this, even were we much more knowledgeable about the micro-ontology I've invoked, it's not obvious that the macro-objects (line drawings) that we treat in a Euclidean manner are well-defined model-theoretical constructions of this underlying physicalist ontology. I allude here to a host of familiar problems, including those about how to individuate macro-objects in terms of subatomic items (of any sort).[3]

---

[3] I've described this last concern in model-theoretical terms. Of course, my own favored syntactic approach holds out even *less* promise in this respect—because what's then required are definitions of macro-terms in the language of fundamental physics.

In any case, it's pretty obvious that the successful application of mathematics to an empirical domain can take place without a definitive characterization of the latter. In light of this, I adopt the deliberately vague word "phenomena" and say that it's phenomena, properties of phenomena, and relations among phenomena that E describes. What a particular chunk of phenomenon turns out to be, ontologically speaking—one thing or many, things with properties they have in themselves, mere (reified) relations to us, even sheer instrumental artifacts—doesn't affect the success of applying mathematics to it.[4]

3. As is well known, the application of mathematics to an empirical domain is often not "exact." This manifests in at least two ways, especially if measurement is involved. First, our ways of instrumentally capturing properties of a domain are always subject to "margins of error"; the mass of Pluto (and its distribution throughout the planet), for example, can be measured only to within a certain degree of exactness, a degree relative to our instrumental capacities at a time—not to mention time and energy. Unsurprisingly, one well-known aspect of science is how instrumental accuracy (and efficiency) increases over time.

But inexactness can arise in a different way: A "triangle" on a piece of paper can be treated as a Euclidean triangle only to within a certain "margin of error" with regard to the width and straightness of its borders. This isn't because of an inability to measure (or draw) the triangle accurately enough; rather, the figure can be treated as a Euclidean triangle only to within a certain "margin of error" because its sides simply *aren't* straight or breadthless: it isn't *that* sort of thing.

This "mismatch" between S and E is handled in applications by allowing the sentences of E, which the theorems of S are mapped to, to be "rough" ones. For purposes of certain applications, Jupiter is "roughly" a point, triangular figures are "roughly" triangles, and so on. Successful mathematical application isn't only unaffected by such "sloppiness" in the sentences of E that we choose to map theorems of S to; it positively flourishes on it: Mathematical theories themselves often enable us to test their applicability to an empirical domain because the deviance, of empirically measured results from mathematically inferred predictions, can be directly traced to the failure of empirical phenomena to conform to mathematical specifications. An easy example: The area rule for rectangles will deviate from the measured area of an empirical rectangle to the extent that the borders of the empirical object aren't straight and sharp and the surface it's on isn't flat.

We can thus explain the failure of a branch of applied mathematics to yield predictions that match empirically derived results to within a certain accuracy. But this point is instructive in another way: Successfully applied

---

[4] Artifactual irregularities due to the dye on a sample or because of the properties of light interacting with a microscope can nevertheless be counted, or their border geometry described and measured.

mathematics doesn't need a "perfect" fit between mathematical language and empirical language (or between mathematical ontology and empirical phenomena). An imperfect fit, where deviance from prediction is explained by deviance of fit, does fine provided, of course, that the deviance from prediction is small enough for whatever our purposes happen to be.

This take on inaccuracy in applied mathematics doesn't require this inaccuracy to be represented in the mapping g itself—g can simply take descriptions of mathematical objects (e.g., triangles) to descriptions of empirical phenomena (e.g., quite crude drawings of triangles)—and we needn't have high expectations that our resulting predictions of the properties of these empirically characterized triangles will be very close to their actual properties (when empirically measured).

Thus, there is a twofold complication in this general story of how mathematics is applied. First, as I mentioned before, E needn't even *be* an accurate or true description of the empirical domain in order for S to successfully apply to it: Often we don't have such a description (the description of the empirical domain, say, is already quite idealized, and it's to this idealized description that S, strictly speaking, is applied). This is one point of describing what E characterizes as *phenomena* rather than as actual items in the empirical domain. But, second, the induced mapping of theorems of S to sentences of E needn't map the theorems of S to the most precise descriptions that E is capable of supplying. This gives us, in principle, two places to look for why deviations from predicted results arise.

4. There is a sense in which the language of E, very often, is already "prepped" for the application of an algorithmic system S to it. Consider the application of Euclidean geometry to figures on a piece of paper, and consider the class of drawings described as "square figures." This is a vague class of drawings, more or less identifiable by eye, and like all empirical categories, there are items that we might be puzzled about how to classify according to it. Consider the word "empirical" just used: Despite the appearance of "square" in the label "square figure," the latter is a normal *empirical* term and not a mathematical term *disguised* as an empirical term. Nor is it an empirical term with a "method of application" via an implicit understanding of a Euclidean term (viz., "square" when it refers to a figure bounded by four equal lines with no width, making four angles with each other of exactly 90 degrees). The latter thought would rapidly force us to conclude that, in fact, there are no "square figures" *really*; it would also leave us wondering how we *understand* how to apply a term when that application presupposes the capacity to recognize such odd objects—can we *see* widthless borders? But this construal of understanding "square figure" is mistaken (and, worse, absurd). There *are* square figures, and everyone is perfectly capable of learning that term so that most of us pick out (pretty much) the same figures.

One piece of evidence for this claim is that geometry, in particular, started out as an empirical study among ancient Egyptians. It's important that the mathematical term "square" can arise by a "mathematization" of an empirical term "square figure," rather than the other way around being the way it *must* go. This point isn't special to Euclidean geometry. Many classifications of empirical phenomena, which apparently use mathematical nomenclature, are actually based on purely empirical terms with their own sets of (empirical) procedures that enable recognition of what those terms pick out. This isn't invariably the case, however, as I'll show.

5. It might seem that because application of an algorithmic system S to an empirical domain E is "prepped" by the language of E, as just described in elucidation (4), we're deprived of the ability to *explain why* a branch of mathematics can be applied empirically. After all, if E already contains such terms as "square figure," "triangular figure," and so on, the possibility of applying Euclidean geometry to E is built into E itself; it's therefore built into the world by the way that E describes that world.

Not every demand for explanation is reasonable. In this particular case, the application of Euclidean geometry to the world—when a description of the world is given at a certain macro-level: a description of the world as we humans see it, feel it, and so on—*needs* no explanation other than that squarish figures, after all, more or less *have* the properties that Euclidean geometry predicts squares to have. Figures drawn, more or less precisely, on flat pieces of paper, act, more or less, like Euclidean figures—there is no more to explain here than that.

The key point is that explanation is always relative to a framework (or level of description) that the explanation is to be mounted *from*. Suppose, therefore, that we take the properties of figures on pieces of papers (as we *do* take them) to be due to the machinations of items of an underlying ontology (i.e., of the various fields of quarks, electrons, etc.). Here what needs explaining is why the mapping of Euclidean language to the empirical macro-language works as well as it does; the story to be told must involve details of the mathematics applied and the underlying ontology of the empirical phenomena it's applied to: why certain idealizations in our description of the empirical domain won't get us into certain sorts of trouble (why certain properties of items in the domain can be disregarded), *and* facts about how those phenomena arise by human intervention. Consider the Euclidean case again. Here we tell a story—about how relatively flat (and hard) surfaces arise—in the language describing the underlying properties of the molecules those surfaces are made of, repellent and attractive forces, and so on. Such a story reveals an actual surface neither very flat nor very hard. But crucial to the applicability of this branch of mathematics to this empirical domain is an idealized description of the level at which human senses and instruments interact with that surface. Relative to the crudeness of our tools and senses, the rich detail of the irregular surface is (literally) overlooked, and this explains the relatively successful application of Euclidean geometry (on the level, anyway, where our senses and instruments are sensitive). *End of elucidations.*

This initial description of the application of mathematics must be modified if it's to capture the most interesting applications of mathematics in the sciences. To motivate the way I do so, notice two points about the language of E (that characterizes an empirical domain) implicitly in place in the examples of line drawings and rigid rods given previously: First, the terms of E describe items we can *observe* and classify by their observed differences. Second, these items are logically independent of the mathematics applied to them: It's not that such items can be (or are) treated as constructions (of some sort) out of the existential commitments of Euclidean geometry (or out of the existential commitments of the number system, in the case of the straight rigid rods).

Neither of these conditions need hold of the terms of E, however. For an example (c) of a case where the second condition doesn't hold, let *Newtonian cohesive-body mathematics* (Ncm, for short) be a family of algorithmic systems that characterize objects moving within a three-dimensional Euclidean space ($\mathcal{R}^3$) over time ($\mathcal{R}$); in choosing $\mathcal{R}^3$ and $\mathcal{R}$, for space and time, respectively, the standard topologies and the standard metrics that can be defined on $\mathcal{R}^3$ and $\mathcal{R}$ are chosen, as well, via a distinguished coordinate system for space and time; this distinguished coordinate system also enables the points in the various $\mathcal{R}^3$ subspaces of $\mathcal{R}^3 \times \mathcal{R}$, namely, $\mathcal{R}^3 \times a$, $a \in \mathcal{R}$, to be identified with each other via their spatial coordinates alone; the same is true of the temporal points (moments). In short, I'm assuming as background geometry for Ncm what's known as "absolute space" and "absolute time." This requires, on the part of Ncm, existential commitments to four-dimensional points, as well as to various set-theoretical objects (including the coordinate system) defined on those points.

It's known that—mathematically speaking—we can get by with less: Euclidean space has a natural metric defined on it, but an *affine manifold* that lacks a metric but still has a privileged set of straight lines suffices for that bit of Newtonian physics that I'll be looking at.[5] This is because Newton's laws of motion, universal gravitation, and the various contact forces assumed, don't distinguish reference frames that differ from each other only with respect to velocity or position.[6]

From some perspectives on how ontology is to be read off of scientific theory, it's very important (ontologically speaking) that Euclidean space not be required as the backdrop for Newtonian physics, because on these views the background space is something that Newtonian physics would otherwise commit us to, even if it brings with it theoretical distinctions (say, regarding absolute position and/or velocity with respect to

---

[5] See, e.g., Friedman 1983, esp. pt. III, or Stein 1967.

[6] A qualification: Velocity-sensitive forces may prove of value within the context of Ncm, but I'm waiving this consideration in the above discussion in order to stress the ontological irrelevance of this issue in any case.

the distinguished coordinates of that space) that can't be observationally determined. From these perspectives on ontology, and Quine's is one of them, any recasting of a theory, even for purely mathematical reasons, is fraught with ontological implications. I argue in chapter 9 that, ontologically speaking, the geometric assumptions adopted as a backdrop for Ncm make no difference at all. So I've adopted the worst case, at least with respect to the issue of our apparent commitments to "absolute space and time"; I intend to show how little effect such mathematical choices really have on such commitments.

The same is true, by the way, even if Newtonian physics is recast more radically—for example, via fields. An important implication of the view urged in this book is that mathematical reconstruals of what is, from the physical point of view, the same theory turn out identical in their ontological commitments. I should add that this is precisely what's to be hoped for.

Also included among the quantifier commitments of Ncm are *moving* point-masses that can have various (fixed) mass values, various initial velocities, and various (initial) locations; the functions that describe the movements and locations of point-masses over time are given by Newton's laws of motion and Newton's universal law of gravitation. One last sort of item is needed: three-dimensional *bodies* that also move in space over time according to Newton's laws of motion and the various force laws applicable to them, Newton's universal law of gravitation among these. These will be called *cohesive bodies*.[7]

Several points should be made about cohesive bodies: First, they're treated via continuum mathematics. This means that although they're composed of points (following Truesdell 1991, hereafter called *material points*), *these* points shouldn't be seen as the point-masses of the last paragraph, if only because, in general, point-masses have nonzero mass but the material points within cohesive bodies generally have zero mass.[8] Indeed, such bodies are highly structured mathematical objects—this facilitates the presentation of continuum mechanics in a mathematically rigorous fashion.[9] Such bodies are taken to deform (in part) because of contact

---

[7] Because point-masses are included among the admissible mathematical objects, "Newtonian cohesive-body mathematics" may sound inaccurate. Regard point-masses as degenerate examples of cohesive bodies.

[8] The mass of a cohesive body is given by a mass density function defined over the volume of that body. See Truesdell 1991 for details.

[9] E.g., see Truesdell and Rajagopal 2000, where we find:

> A body β is a set that has a topological structure and a measure structure. It is assumed to be a σ-finite measure space with a nonnegative measure $\mu(\mathcal{P})$ defined over a σ-ring of subsets $\mathcal{P}$ of β called subparts of the body. The open sets of β are assumed to be the σ-ring of sets. The members of the smallest σ-ring containing the open sets are called Borel sets of β. (p. 1)

Also see Truesdell 1991, chap. 1.

and body forces;[10] they're also taken to have *constitutive relations*, which govern how such bodies move and how they move and deform under the effects of such forces.

Two points: First, we may discover that the empirical objects that rational continuum mechanics successfully describes are ones to which constitutive relations apply successfully (within a certain degree of accuracy) because what corresponds empirically to these constitutive relations can be described as various forces between, say, the molecules that such empirical bodies are composed of and, ultimately, the forces between *subatomic* particles. But *this* study—with its ultimate revelation of other force laws in nature—necessarily moves beyond the Newtonian framework.[11] For my purposes, however, and staying within Ncm, we can still regard constitutive relations as corresponding to (idealized) descriptions of the materials that the empirical bodies (corresponding to cohesive bodies) are composed of. In turn, composition out of these different materials can be seen to induce various kinds of (resistive) (internal) forces.[12] Second, in claiming this, by no means am I claiming that a "reduction" of continuum mechanics to the (quantum) physics of subatomic particles is in the offing, or that, therefore, in some sense, a reduction of continuum mechanics to the physics of point-masses, say, is also in the offing. Apart from the fact that subatomic particles aren't point-masses, reductions in this broad sense are extremely rare in the sciences.

Has a family of algorithmic system been specified by the foregoing? A family of algorithmic systems as Euclidean geometry and PA are understood to be such systems? Well, apart from my being rather vague about the specific axioms supposedly involved here, the answer is: Of course. There is no restriction on which concepts can occur in a family of algorithmic systems (no concept is "purely mathematical"). And just as group theory has many models amenable to study, and classes of which can be singled out by additional axioms, so too Ncm has many models or collections of models that can be similarly studied. The mathematical nature of the latter subject matter is shown by the absence of concern over whether any of these models is empirically "right"; *that* concern is to be raised in

---

[10] Body forces are related to the mass density of such bodies taken over their volume. Contact forces are related to the density taken over the body's surface area.

[11] It needn't have, of course: Empirical results forced us beyond something compatible with the Newtonian framework to quantum mechanics.

[12] Not adopting this last move leads to a terminological denial of Newton's first law. See Truesdell 1991, p. 66. For my purposes, it really doesn't much matter how constitutive relations are understood because I argue in chapter 9 that, just like forces, they dissolve, in any case, into sets of mathematical constraints on possible motions of the material points contained in cohesive bodies—*not*, i.e., into something we need take ontological account of (unless we *desert* the context of Ncm).

exactly the same way as the success of the application of Euclidean geometry was raised earlier in this chapter.[13]

Now imagine an *empirical world* a bit more dreary than ours: There are just (fairly) arbitrarily sized (and shaped) cohesive *lumps*, made up of various materials, some more rigid than others, that travel endlessly through space, some deforming, some not, and that occasionally break apart into smaller chunks or deform under gravitational forces (e.g., during rotation) or when they come in contact with one another. (They can also join up, of course.) Instead of what empirically corresponds to electromagnetism, weak and strong forces, and the zoo of fields of subatomic particles with their complicated laws that govern our world, there are just what empirically corresponds to the concepts in Ncm of various contact forces and one body force (gravity). These lumps are also made up of different materials that affect how they react to various external and internal forces, and these correspond to various constitutive relations in Ncm. The empirical language E contains names for some of those lumps and some general predicates and relations that hold of them: "x is spherical," "x is more massive than y," "x has more volume than y," "x is farther away from y than w is from z," and so on.

The quantifier commitments of E are all and only such lumps. Ncm is applied directly to the lumps in a way similar to how Euclidean geometry is applied to figures on paper: Find a useful number of categories of cohesive bodies, composed of (idealized) kinds of materials, for example, that we can map classes of lumps to, and then subsequently describe the kinematic and dynamic properties of these classes.[14]

Applying Ncm allows us to characterize the properties of such lumps by approximation methods, treating them as (a class of) cohesive bodies to within a certain margin of error: approximating the amount and distribution of the mass in such lumps, their volume and surface areas, and so on, by mathematically manageable functions of various sorts,[15] the ways

---

[13] For explicit axiomatizations of (some of) analytical mechanics and (some of) continuum mechanics, see Noll 1974; for references to other axiomatizations, and for some humorous, and yet nevertheless nasty, polemics, see Truesdell 1980–81.

[14] In Truesdell 1991, we find: "The approach is like that of Euclidean geometry, in which, after statement of the axioms satisfied by all geometric objects, theorems characterizing and relating classes of figures are proved" (p. 200). He then adds: "Since mechanics is a discipline vastly more subtle and sensitive than geometry, the parallel stops here and does not extend to the theorems themselves or even to the methods of constructing proofs." As an illustration of this, consider Truesdell's 1991 gloss on constitutive relations: "[Constitutive relations] define *materials*, which are mathematical idealizations of the materials encountered in nature.... Typical constitutive classes are the bodies called rigid or solid or fluid, isotropic or anisotropic" (p. 6).

[15] "Mathematically manageable" includes, but isn't restricted to, *smooth*, i.e., differentiable as many times as needed.

that such lumps perform when acted on by external forces, various ways in which the parts of the lump deform and move, by idealized constitutive relations, and so on.

An easy example: If a lump is sufficiently far from the other lumps, we can approximate its gravitational impact on the other lumps by *ignoring* nearly everything about it, assuming, that is, that all its mass is concentrated at its center of mass. If, however, that lump strikes another lump, we'll have to take internal properties into account—in what detail and how, what can still be ignored, and what not, is empirically determined by the behavior of the lumps under impact.

We can imagine an easy epistemic generalization of this case (d), where the lumps recognized and categorized in the language of E (via observation) are instead so recognized and categorized (at least some of the time) by instrumentation of some sort. In this generalization, we're still taken to have, that is, thick epistemic access to the objects of E, and we recognize the properties of such objects via that access.

For example, consider certain properties of the planets and other bodies in our solar system. We have substantial access to these properties by means of various instruments, not by naked sensory observation, and the kinds of instrumental access available enable us to determine these properties in great detail. This access, however, allows us to apply empirical terms to these bodies, where such terms are already active in ordinary life.[16]

In both examples (c) and (d), although E has its own set of terms for the lumps and their properties that it allows us to talk about, and although there are methods for gaining access to the (empirical) properties and relations (characterized by E) that lumps have, to apply Ncm to the existential commitments of E we *reconstrue* those commitments as quantifier commitments of Ncm (reconstrue lumps as cohesive bodies) and reconstrue the empirical properties and relations of lumps as the mathematized properties and relations of those cohesive bodies.

There are two ways to ring changes on the Ncm examples: The first (e) is if there are properties of lumps their reconstrual in Ncm leaves out *altogether* (and not merely because their effects are minimal for purposes of certain applications of Ncm). For example, such lumps might be colored in various ways without the character of this coloring being amenable to an explanation in terms of a physics based in Ncm. In this case, the original model for the application of mathematics still applies: There is a mapping of the language of Ncm to that of E that maps cohesive bodies to lumps, and maps $\Re^3$ to nothing at all. By the way, I *don't* mean that physical space is a void devoid of properties and that $\Re^3$ is mapped to

---

[16] E.g., we have good descriptions of the rough volumes of many of these bodies, and their makeup—how much water is present, what gases their atmospheres are composed of (if they've got atmospheres), the temperature variations on the surfaces of such bodies (if they've got well-defined surfaces), etc.

that; I mean that $\Re^3$ isn't *mapped* to anything at all. I say more about this in chapter 9, in the section titled "Absolute Space and Time."

The second possibility (f) is where we have no antecedent language *E* of lumps to start from. Imagine, instead, a domain to which we have only instrumental access (not observational access), and it's discovered that Ncm applies to the empirical patter that our instruments detect. Suppose we find that certain cohesive bodies—and notice I'm now using the *mathematical* language of Ncm to describe the situation—have detectable properties (their "locations," their "velocities," and their "distances" from each other, say),[17] and we're able to predict their future relations and properties on the basis of initial data our instruments discover about them plus the application of Ncm. Imagine also that the language used is the pure language of Ncm and that we find that such mathematical properties as location, velocity, and so on, can be correlated with instrumental detections. In this last case, no antecedent *E* is to be had. And, as footnote 17 indicated, this is no mere thought experiment: Exactly this kind of situation obtains with the physics of subatomic particles and their properties.

It's worth comparing this case with a previous one. In example (d), although instrumental access is as essential to our determination of the properties of the objects under scrutiny as it is here, that instrumental access is to macro-properties to which we have (more or less) observational access in other contexts. The recent discovery of substantial amounts of water on Mars, for example, discovered by purely instrumental means, is a discovery that may well be exploited in the near future by Martian colonizers who will just *see* the stuff.

But this is false of the properties in the case under current consideration: *These* properties that instruments enable us to access are properties we *never* have observational access to. This difference is important because in the first case the (rough) scopes of the predicates denoting the properties in question are indicated not just by our instrumental access to their presences, but by our observational access to their presences, as well.

Why does this matter? I'll say more later[18] but, roughly, the point is this: We generally don't allow an instrument (or observational means, for that matter) to *define* the extension of a noun phrase or predicate (to pick out a kind of thing or a property). But the role of *instrumentation* with

---

[17] Why the scare quotes? Well, what could happen is similar to what *has* happened with respect to the properties of subatomic particles. "Spin," e.g., although analogous to "spin" as applied to macro-objects like tops, *can't* be the same thing. We might similarly know that, in fact, the objects under study can't be massive lumps careening through space with *real* "locations," "velocities," etc., because the so-called "lumps" are items detected via instruments in a context that excludes this interpretation; e.g., they correspond to *intensive* properties of a phenomenon and aren't distributed (in space) at all. Nevertheless, we might still find Ncm empirically adequate; i.e., we might find that what it predicts is empirically verified to whatever degree of preciseness we're capable of testing it to.

[18] Actually, I've said quite a lot already. See my 2000b, pts. III and IV.

respect to the noun phrases and predicates of a language is usually even more restricted than this: We don't allow instrumental access to something to contribute significantly to how we project a predicate or noun phrase over that (kind of) somethings. This is for an empirical reason: Instrumentation, even the most sophisticated and powerful, is always extremely narrow in scope, and the properties of an object being instrumentally accessed, even those very properties being accessed, are understood to far outstrip our instrumental capacity to study them.[19] But observation, in general, provides a broad enough, and strong enough, access to objects and their properties—those we have observational access to, anyway—that we find such observational access (if robust enough—not furtive glances, as it were) sufficient to *support* the reference of a term (referring to or holding of that something), that is, as sufficient for the substantial projection of a predicate or noun phrase.

This does *not* mean that there are terms with extensions restricted to observables—on the contrary. But it does mean that, because the class of kinds of objects to which we have observational access is broad and substantial enough, because the number of *different* properties (accessible by observation) of such objects is so great, and finally, because observation provides enough of an epistemic surety (not, of course, infallibility) of what's been observed—an epistemic surety *as* reliable as *any* method of epistemic access to things that we currently have—sheer observation is seen as providing enough of a basis for the application of a term that we're comfortable coining terms that refer (*among* other things) to what's observed and to observed properties of such things.

This isn't the case with instrumental access to objects and their properties, when such access is *not* linked in some way to observational properties and things (as with water on Mars).[20] In such cases, something else is needed to supply the (provisional) referential basis for the application of such terms, and what's needed is supplied by *mathematical nomenclature* instead, because mathematical nomenclature brings with it mathematical *truths*, and it's the stability and wide scope of these truths that enable us to project the extensions of theoretical terms beyond the instrumental access we have to items to which such terms are taken to refer.

As a result, the first model for the application of mathematics is useless; there are no terms of an empirical language E that we can map the

---

[19] In describing this as empirical, I'm explicitly allowing that it could change.

[20] "In some way" is vague. I'm alluding to my discussion of ob- and ob*-similar sorts found in my 2000b, pt. II, § 3. My purpose in so alluding is to forestall the thought that the above contrast is a simple one where either instrumental access is to something that is otherwise observable, or it's *not*. Things are, as always, more complicated. Many of the instrumental ways we have of determining the presence of various atmospheric gases, for example, are linked "in some way" to observation, and linked so that we're secure about coining (projectable) terms that refer to such gases. But this isn't because these gases are straightforwardly *observable*.

algorithmic system *to*. Instead, some of the existential commitments of the algorithmic system have mathematical properties and relations that are *correlated* with the properties and relations of *something* empirical, which, as it turns out, *lacks* an antecedent characterization but which can nevertheless be detected instrumentally. That is, we have applied a mathematical system to phenomena by simply forging thick epistemic access to items that are otherwise picked out by purely mathematical nomenclature. This *doesn't* mean we've (somehow) forged thick epistemic access to mathematical objects; rather, it means that (some of) the mathematical posits are proxying for something empirical that we can't otherwise describe: Pure mathematical terminology has been drafted as a system of characterization for something empirical.[21]

If this way of speaking seems ontologically offensive ("mathematical *posits* are proxying for . . . "), the point can be put this way instead: Mathematical terminology, because of its empirical fit to something, has been co-opted to describe that something. Once such terminology is in place in an empirical theory, we can then—as we would with anything empirical—root it to the world by forging thick epistemic access to (some of) the posits.

As I show explicitly in chapter 9, but as the reader can already detect from the foregoing, the way of applying mathematics exemplified by example (f) has important implications. I conclude this chapter by first raising one important implication, and motivating by means of it the project of chapter 9; and then I deflect a different purported implication that has been drawn from examples like (f)—namely, the claim that we have no (epistemic) justification for introducing mathematical nomenclature into physical theories, the way I described (f) as doing, and letting that nomenclature dictate how terms are to refer *empirically*; when mathematical nomenclature has a nonempirical origin—as it so often does—there is no epistemic justification for thinking that it will fit anything in the empirical world.

First, an implication I take seriously: Because in cases like (f) all the nomenclature—the predicates and constants—are imported from pure mathematics, there will be, in general, no way of separating out (pure) mathematical terminology from nonmathematical terminology as those nominalistically inclined philosophers need who are committed to the Quine-Putnam indispensability thesis (plus Quine's criterion for ontological commitment), and who are also intent on eliminating ontological commitments from the language of a physical theory. Any such nominalist, who desires to *extract* the "nominalist" sentential content from mathematized physical theories, or who even just wants to "postulate" the

---

[21] But it mustn't be forgotten that this does *not* mean that *all* of the distinctions the applied mathematical terminology brings with it have been fitted with this empirical role. On the contrary, as I show in chapter 9.

existence of such content, and wants to claim that such nominalist contents are true and the rest of the mathematical accompaniments are *not*, faces the considerations raised in chapters 1 and 2.[22]

Nevertheless, the position on ontology developed in part I, and in chapters 6 and 7 of part II, has resources for distinguishing ultrathin posits from thick or thin posits *even in* these cases, and consequently, such a position has resources for making pertinent ontological distinctions (between which posits in such a physical theory should be taken to exist and which not) *even though* the language of that physical theory is through-and-through mathematical, and even though, therefore, the (physical) theories in question must quantify over ultrathin posits. I illustrate this project explicitly in chapter 9 with respect to examples (a) – (f) described in this chapter.

Next, I note again, in order to set up the concern with the success of applied mathematics, that the kind-terms, the "projectable predicates," to use deliberately Goodmanian language, must be borrowed directly in case (f)—and the numerous other cases like (f)—from (pure) mathematics. This may not seem to be the case in a more general sense: Ncm, after all, uses notions such as "cohesive body," "velocity," and so on, which seem borrowed from the empirical world, even if they have been modified severely ("idealized," as it's put) in the process of that borrowing. Indeed, it may seem clear that all the notions of Ncm, including that of "space-time point," arise fairly directly from empirical notions in this way. So although the application of Ncm in case (f) is to an empirical arena from which the notions of Ncm did not originally arise, still, the notions (ultimately) have an empirical source.[23]

But this needn't be the case. Quantum mechanics, for example, uses the notion of a complex Hilbert space, where the vectors in question are complex wave functions. These complex-valued functions, although a crucial part of the quantum mechanical formalism, aren't themselves directly given a physical interpretation. Rather, it's the results of various mathematical operations *on* such wave functions (e.g., integrating the product of the wave function, its complex conjugate, and some other function—the latter depending on what, precisely, is being measured) that yields, in the standard interpretation of this formalism, *probabilities* of position, momentum, and so on, for particles. The *latter* are called, for example, by Landau and Lifshitz (1977), "physical quantities."

---

[22] Such a nominalist faces another problem to be raised in chapter 9: Any restriction of the predicates and noun phrases to what's taken to be real results in an intractable theory: In order to be able to *say* what's true about what's real, we have say things about *more* (our terms have to refer to more) than what's real.

[23] I'm going along with this view only for the sake of argument: Long familiarity with points makes us think of them differently from how we think of, say, the complex wave functions in a Hilbert space that I'll mention in a moment. But I don't think that points, really, are any more empirical looking than are complex wave functions. I don't, e.g., (really) know how to visualize one any more than the other.

At least one philosopher—Mark Steiner (1998, and earlier, 1989)—has explicitly discussed this role of mathematical terminology in modern science, and he has raised the concern that the empirical *success* of the drafting of mathematical kind-terms for the role of projectable (empirical) predicates in empirical theories involves an unexplained methodological miracle. The worry is that mathematical systems, designed as they are to be implementable by mathematicians (to be elegant, aesthetically appealing, easy—relatively speaking—to prove theorems in, etc.), are thus parochial enough in the desiderata for their construction to raise the worry that such systems shouldn't be empirically successful (any more than any other arbitrary set of rules for manipulating linguistic tokens should be, if such a set is designed with our own psychological proclivities squarely in mind).

The concern divides neatly into two distinct worries that we should *keep* distinct (esp. because the ways of soothing the worries aren't the same). The first worry arises from the request for an explanation for *why* the empirical applications of such mathematical systems are successful. I've discussed this worry at length elsewhere,[24] but what's called for are the sorts of considerations I raised earlier in this chapter about how we explain the applicability of mathematics; once, that is, we have an underlying theory of something in place, we can *explain* why the mathematics applied to *it* works (e.g., tabletops and Euclidean geometry). However, sometimes we must simply acquiesce in the nonexplanatory bruteness of the success of our mathematics, because an explanation of the success of an application of a mathematical theory, and an explanation for the success of an empirical *scientific* theory, for that matter, is something—in this sense of "explanation"—that's only available from the vantage point of *another* theory. We can explain the success of Newtonian physics—its relative accuracy—from the vantage point of relativity theory, and some day we may explain the success of the application of the non-Euclidean geometry used in general relativity from the vantage point of, say, string theory. But until and unless we have another theory in place to compare general relativity theory *to*, we can't explain its success apart from simply uttering the banal truism that it's *true*.

There is, as I mentioned, a second worry that Steiner is explicitly concerned with:[25] whether an epistemic *justification* for engaging in this practice of drafting mathematical terminology for empirical purposes *can be given at all*. This worry is distinguishable from the first worry like so: Imagine we're considering what *sort* of theory we should construct and what tools we should use in doing so. Steiner's worry is that many of the tools used by contemporary physicists in constructing such theories *can't* be given an adequate *epistemic rationale*. That is, it can't be explained why

---

[24] See my 2000a.

[25] Two points: First, I did *not* respond to this second worry in my 2000a; I'm doing so now. Second, Steiner (1998) doesn't always clearly distinguish the two worries.

the practices of theory construction in contemporary physics are reasonable ones to engage in. Notice that *this* concern won't be satisfied by merely invoking the incredible empirical success of the strategies in question, for there is the fear that dumb luck has cheerfully been on our side in our theory-constructing endeavors.[26] Nor is it to be satisfied by the sort of balm that soothes the first worry: Explaining why the tricks worked in hindsight *won't* explain why it was sensible to use those tricks in the *first* place. No, only a (general) explanation of why the strategies (that modern physicists employed to design their theories) were reasonable strategies to employ *despite our not knowing if they would succeed* will soothe this worry.[27]

But *does* the way that (contemporary) physicists design physical theories make it an epistemic miracle that such theories turn out to be empirically adequate?[28] Are their strategies really so unreasonable (epistemically speaking)? Steiner thinks so, because such theories, housed as they are in pure mathematical nomenclature, are developed by the use of what he describes as "Pythagorean" and "formalist" analogies.

"By a 'Pythagorean' analogy or taxonomy at time t," Steiner writes (1998, p. 54), "I mean a mathematical analogy between physical laws (or other descriptions) not paraphrasable at t into nonmathematical language."[29] And on the same page we find: "By a 'formalist' analogy or taxonomy, I mean one based on the syntax or even orthography of the language or notation of physical theories, rather than what (if anything) it expresses."

I take up Pythagorean analogies first and turn to formalist analogies after. Steiner's calling something a "Pythagorean analogy" is meant to

---

[26] Steiner (1998, p. 10), who deliberately hints of his inclination toward theism, is therefore inclined to think that "dumb luck" in the form of divine kindliness *is* involved. Notice that the "dumb luck" worry undercuts an inductive justification for *continuing* to use what we'd otherwise regard as unjustifiable strategies for designing empirically adequate theories: Why assume that such luck will continue? (Think of Job, and what happened to *him*.)

[27] Of course, there is no escaping the fact that, in certain cases, success *is* a matter of dumb luck, *even in* intellectual endeavors. But that can't *generally* be the case with respect to theory-constructing strategies, can it?

[28] Steiner (1998) doesn't couch the issue this way, but he's out to show that scientific methodology is "nonnaturalistic" and thus presupposes a "user-friendly" universe. That's equivalent to the concern as it's put here.

[29] This definition is unfortunate in seeming to imply that if we *can't* describe an empirical domain in *purely* nonmathematical language, and therefore, if we can't give physical motivations for a theory in purely nonmathematical language, then (by definition) *any* considerations about the properties of that domain will be "Pythagorean." But Steiner's examples make clear that he doesn't want to press the Pythagorean charge so cheaply. He is concerned with cases, as his examples show, where a theory is *modified* on grounds connected to "internal" mathematical values (ease of computation, aesthetic beauty, etc.). The concern thus turns out to be similar to the traditional charge leveled at the coherentist theory of truth: *Why* should (internal) considerations of simplicity, coherence, etc., suffice to make a theory *true* of what's *independent* of us? The internal considerations are parochial, but what the theory is supposedly about is *not*. (Also see chapter 4.)

draw attention to the basis of such analogies in *mathematical* similarities, not on what we'd describe as physical similarities. So, for example, applying similar laws to both airflow and water flow turns on the physical similarity of the phenomena those laws are applied to. However, the strategy quoted from Purcell in chapter 2 turns on no physical analogy between the microstructure of a magnetic substance and that of an item composed of tiny conducting ribbons, but only on the fact that (at a certain distance) the fields of both are numerically similar. So too, there is the analogy Cartwright (1983) mentions at work when the harmonic oscillator model is used in cases in quantum mechanics "even when it is difficult to figure out exactly what is supposed to be oscillating" (p. 145).

Steiner describes several examples of what he calls Pythagorean strategies, where modifications are made either in physically applied theories on the basis of suggestive mathematics alone, or where a mathematized theory is applied to a fresh empirical domain without *physical* justification. I grant, for the sake of argument, that these cases are largely as Steiner describes them. He thinks that in such cases there is no more *epistemic* justification for modifying the structure of a mathematized theory on purely mathematical grounds in order to make it empirically adequate than there is for modifying the rules of a game, like chess, to see if the result of doing *that* yields a theory that will be empirically adequate.[30]

---

[30] See Steiner (1998, chap. 4). In giving these examples, he notes how often intuitions of elegance, and other parochial considerations about theories, are used by physicists as indicators of the modifications being in the right direction, empirically speaking. Steiner's point, I want to stress again, is not that such modifications don't lead to empirical success, nor is it that such successes can't be physically explained after the fact. His claim is that the procedure makes no sense to begin with: Epistemically, we've no reason to think that modifications in theories due to notational analogies or mathematical elegance will lead to something empirically adequate. As I said, I'll assume for the sake of argument that his examples pretty much illustrate his claim that such (purely mathematical) analogies are operating. (See the quote from Yang in Steiner 1998, p. 70; the discussion of how Maxwell modified Ampère's law, pp. 77–80; Schrödinger's equation on pp. 81–2, etc.) However, one must be more careful than Steiner sometimes is about whether or not a purely mathematical distinction has physical content. Steiner (1998) writes: "It is convenient—and therefore customary—to use the concept of the Taylor series in classifying phenomena as 'first-order', 'second-order', etc. That is, suppose we have a function that we feel tells the 'whole truth' about nature. If we expand it as a series, we can say what nature would be like if we ignore terms beyond the first-order, second-order, etc., terms" (p. 69). He then claims in the next paragraph:

> The classification of *phenomena* (as distinct from terms) as first-order, second-order, etc., is certainly not one that derives from observation, but convenience. We could just as well have classified phenomena according to decimal powers: i.e., magnitudes from 0 to 9 belong in one class, from 10 to 99 in the next class, etc. The real distinction is that between first-order, second-order, . . . *terms*—a mathematical distinction deriving from the calculus of Taylor series. Projecting the distinction on the phenomena is anthropocentric, but scientists do it anyhow.
>
> Anyone who has studied the growth of functions realizes that if an empirical phenomenon can be captured by a function expandable in a Taylor series (with a number of nonzero coefficients, of course), then there can be practically no doubt that interestingly distinguish-

Is the practice of using purely mathematical analogies to construct new physical theories from old ones a practice without epistemic justification? No. The mistake in thinking otherwise is one Goodman saw through *long ago*. By the time we get (scientifically speaking) to arcane empirical domains (objects excitedly zipping along near light speed, strings seductively wriggling in multiple dimensions), we *already own* a sophisticated and substantial mathematical physics that works empirically quite successfully over a large domain. (This is true, e.g., at the onset of the discovery of both special and general relativity, as well as at the onset of the discovery of quantum mechanics.)

The empirical success of our already mathematized theories means that we have good *epistemic* reasons to tinker with them (mathematically) to see if something empirically still more adequate can emerge. This *even* makes sense in those cases—which are legion, as Steiner aptly illustrates—when the physical justification for the application or extension of a mathematized theory is *absent*: If certain empirical measurements have a *form* suggestive of a certain mathematical nomenclature, it makes good epistemic sense to *apply* that mathematical nomenclature to the case "by analogy": Doing so might work. It even makes sense in cases where in applying a mathematized theory to a new domain we shift on what parts of that theory we take to be physically significant. This is especially the case if there is no clear physical story in place for why the mathematized theory in question should work in the new domain.[31]

Of course, none of this need work out at all (and often doesn't). But, if a well-developed theory is already in place, tinkering with it is the obvious place to begin. And, again, it's worth stressing that we might later gain a physical understanding of why the mathematics worked (as well as it did) or we might not: The point is only that Steiner's suggestion that going down this road to begin with is epistemically "dotty"—as epistemically justified as, say, the application of numerology to politics—is off the mark.

What about what Steiner calls "formalist analogies"? Steiner's definition (given earlier) doesn't make this clear, but his examples show that

---

able physical processes are at work, because the different terms in a Taylor series grow at (exponentially) different rates. Such is hardly true of a classification of phenomena according to decimal powers.

[31] Steiner (1998, pp. 82–3) mentions cases where previously physically unreal solutions to equations are reassessed: Such gives rise to the Dirac's prediction of the positron and Finkelstein's black holes, for eample. Consider the following easy analogy: Imagine a rectangle with two sides, $x-3$ and $x+1$, with area 21. Solving this problem in the obvious way gives two solutions: $x=6$ and $x=-4$, only the first of which is "physically real." Imagine, however, that we find a new empirical domain to apply "areas" to, where "areas" in this domain are instrumentally detected in a certain way and are *not* spatial in nature. Given that the same simple computational facts about areas apply to "areas," and given that we can't tell a physical story for why the area rule applies to "areas," why is it epistemically unjustified to explore the possibility that the "physically unreal" answer has a role to play in this new empirical domain?

his concern is with the easy and facile way physicists often have with a mathematical formalism, manipulating and exploiting it in various ways without possessing (sometimes *any*) mathematical justification for doing so—the reasoning isn't "mathematically rigorous," as it's often put.[32] In these cases, what one should be struck by isn't that the resulting "mathematics" is physically applicable but that the resulting "mathematics" can subsequently be made into cogent *genuine* mathematics, because once an epistemic justification is in place for tinkering with a mathematized physical theory (in order to produce something more empirically adequate), there is certainly plenty of epistemic justification left for using proof-theoretic *shortcuts* and strategies in that process. Consider the Dirac delta-function: That such a thing was useful to manipulate formulas in quantum mechanics was obvious (i.e., its *empirical adequacy* was obvious). That it could be embedded in a genuine mathematical theory in a consistent way was not. So, of course, the facile way physicists sometimes have with mathematics poses a danger: An approach may subsequently prove mathematically incoherent. But *this* is a risk that all parties concerned are well aware of—and it's not *enough* of an epistemic danger to make facile mathematizing something to be avoided.

Let me close this chapter with one last point about this matter: Steiner claims that the methodological practice of treating, *where possible*, refuted laws (or theories, for that matter) as special or limiting cases of new laws (or theories) is "Pythagorean." What Steiner means by this, as he usually does, is that there is no antecedent epistemic justification for doing so. The methodology, at least insofar as the aim is to generate an empirically adequate theory, is no better than flipping coins.

You might have thought that the justification for this practice is *easy* to establish: The problem with physical theories, always, is that it's never easy to apply them empirically. What's required, always, are intricate *theory-specific* applications of mathematical simplifications and idealizations. That is to say, it's not the case, when faced with a radically new theory, and the question of how to apply it or to extract implications from it, that one has a bag of tricks already in hand that tames the mathematics of that theory. Now Steiner seems to notice just this point,[33] but then he writes: "[I]sn't it just dogmatism (and certainly anthropocentric) to assume that *nature* cares how much time we have invested learning mathematical methods?" (p. 107). This misses the point. In deciding how to modify refuted theories or how to generalize theories so that they have greater scope, we only need epistemic justification for starting, as it were, *here* rather than *there*. And certainly, that a particular modification of an

---

[32] See, e.g., Steiner 1998, pp. 151, 161, 176.

[33] Steiner (1998) writes: "Learning a physical theory means, above all, learning the 'mathematical methods' appropriate for the theory. This includes tricks for solving or approximating solutions for equations, and many other things. If a new theory yields, mathematically, the old, then all of those mathematical techniques are still useful" (p. 107).

already well-understood theory will result in a theory that, relatively speaking, is also well understood is a good reason to *start* by so modifying a theory we already know how to apply in a given empirical domain. Later, after such a strategy has failed, we're epistemically justified in considering something more drastic.

It's also worth noting that we're often forced to tinker with such a theory precisely because we've expanded the phenomenal domain of application of that theory and want to redesign the theory to suit. But that the original theory fit the original domain so well provides evidence—especially if the *expansion* of the domain of application is related to the original domain in specific ways—that a modification of the original theory may be all that's needed.

9

Applied Mathematics and Ontology

Given the ontological conclusions about ultrathin, thin, and thick posits presented in chapter 7, the job at hand looks straightforward: We examine examples (a) – (f) of applied mathematics (from chapter 8) and sort the posits into the three categories. Nominalists then cheerfully commit themselves only to the thick and thin posits. This glib summary is pretty much right—the foregoing apparatus determines what we're committed to and what not when applying mathematics. Still, the route from theory to application is never straightforward, not even in philosophy: Some of this will be rough going because it's sometimes controversial which posits are which.

I start (naturally) with the easiest cases: The ontological commitments that arise in these examples, those of (a) figures on pieces of paper and (b) rigid rods, are easy to determine. We're only committed to what the respective languages E commit us to: straight rigid rods in the one case, and figures on pieces of paper in the other—for, presumably, we have thick epistemic access to both these sorts of items and take them to be ontologically independent. The applied mathematics is a simple application of a pure branch of mathematics, where an explanation of its empirical success doesn't require taking any of its posits as other than ultrathin. (The neatness of the ontological lessons of (a) and (b) is due to the neatness of how the mathematics is applied to an empirical domain in these cases: The language of E is logically independent of the mathematics applied to it.)

Most of this chapter addresses the first *Newtonian cohesive-body mathematics* (Ncm) example, (c), where classical continuum and point-mass mechanics are applied to lumps careening through space. Doing this may seem to involve substantial ontological commitments apart from the lumps of E, both because of how the mathematics is applied, and also because of the physical laws themselves. I (roughly) state my position on its ontology to start, and then explain (rather at length, I'm afraid) why the other things one might naturally think the example commits us to are ultrathin.

The first Ncm example (c) is, ontologically, on a par with the (a) and (b) figures and rods examples: We're not committed to anything more than E's commitments—lumps of matter. That the applying of Ncm to E is implemented by a reconstrual of lumps as *cohesive bodies* containing material points doesn't require there to *really* be such cohesive bodies containing material *points* or, in any case, there to *really* be material points. The same is true of point-masses. Similarly, the need to use a coordinate system of geometric points wherein the locations, velocities, and so on, of these lumps are plotted doesn't ontologically commit us to the existence of any sort of purely geometric object—neither points, nor the various sets of such points, nor the other two- or three-dimensional geometric objects. Nor does it commit us to "absolute space," a substantial something or other that lumps travel through. Finally, despite the presence of Newton's law of universal gravitation and Newton's laws of motion, as well as the other force laws that we apply to E, we're not even committed to *forces*: neither body forces (e.g., gravitation) nor contact forces. (In short, the situation is as ontologically meager as any hopeful nominalist will want.)[1]

A word about the argumentative strategy of this chapter: I first approach denying an ontological commitment to forces via an avoidance of *terms* referring to such forces. This (linguistic) approach to ontology—although quite widespread among philosophers—proves unsatisfactory. I then switch to an approach motivated by the discussion in part I and in chapters 6 and 7.

Forces

I suggested in chapter 1 that the appearance of a schematic letter apparently standing for forces[2] in Newton's second law, $\mathbf{F} = m\mathbf{a}$, needn't indicate that linguistic items referring to forces occur in the sentences that this schema stands for. Imagine first that instead of extended lumps there are only a finite number of point-masses that the terms of E refer to (and

---

[1] Nor can constitutive relations be construed as indicating commitments. Of course, all this changes when we shift theories *and* shift our thick epistemic access to include the internal constituents of lumps. But that's another matter.

[2] Although I primarily focus on forces in what follows, it's easy to see that what I claim, if right, applies to torques. It also applies to constitutive relations for the same reasons—I say a little about this shortly.

that move about). In this case, instantiating Newton's laws of motion and Newton's universal law of gravitation to this model results in differential equations where *no* term for any force appears.[3]

Given the assumptions (1) that the ontological commitments of a theory are all *among* the quantifier commitments of a theory (and distinguished from the other quantifier commitments by a special existence predicate) and (2) that the appearance of *force terms* in the laws is as *schematic letters*, one presumes to deny ontological commitments to forces. It's true that the differential equations describing the motions of the point-masses are sensitive to pairwise relations *between* the point-masses: their distances from each other and the amount of mass each point-mass carries. To say this much is to recapitulate how Newton's schematic law of gravitation and his schematic laws of motion *codify* the differential equations governing this situation: There seems no point at which a commitment to *forces* is made.

I shortly consider the situation if the schematic status of terms apparently designating forces is denied. I should add, to avoid misunderstanding, that in any case I'm *open* to the view that quantifier commitment in a first-order language isn't a necessary condition on ontological commitment to something: In principle, there could be ontologically independent items that we have thick epistemic access to but that are best codified, in a first-order theory, as *predicates*. In this case, I'd describe such items as thick posits, but contrary to how the terminology of such posits was set up in chapters 6 and 7, terms denoting them wouldn't appear among the quantifier commitments of the theory. I didn't frame my discussion in those chapters to take account of this possibility,[4] but for purposes of analyzing the ontological commitments of a scientific theory (as applied), it's important to at least *consider* the possibility that genuine commitments to items we methodologically treat as ontologically independent of us are tucked notationally into the predicates of a theory as well as into the noun phrases. I discuss this further shortly.

Let's return to our discussion of forces. Fundamental forces, of course, aren't restricted to gravity. One point of schematizing Newton's second law is to allow it to apply to *any* force. But this doesn't affect the foregoing considerations (as long as we remain within the context of applying

---

[3] There are constants for the masses carried by each point-mass, for the initial values of the distances between point-masses, and there is also a gravitational constant. There are explicit time-dependent acceleration variables. Relative to the distinguished coordinate system for absolute space and time, initial values for locations and velocities for those point-masses at a particular time also appear, and by virtue of the mathematics of the calculus, implicit time-dependent variables for velocity and position are present. *That's it.* In Quinean terms, quantification occurs over mathematical quantities, a physical constant, and several variables concerning space and time. There is *no* room in any of this (first-order) language, *logically* speaking, for terms that pick out forces.

[4] It wouldn't be hard to do so—but it seemed unnecessary given the arguments to be established in chapters 6 and 7.

Ncm) because just as gravity, as a force, is only a way of characterizing *some* of the laws governing the *movements* of objects, and not an actual item (a "force") quantified over by the sentences of Newtonian physics, so too, the presence of these other purported forces comes to the same thing: They don't emerge, when law-schemas are instantiated, as items quantified over. All forces, as they appear in law-schemas, are indicators of what the sentences describing the motions of objects of specific sorts under specific initial conditions must look like.[5]

Newton's laws aren't restricted in application to cases where the effects of "fundamental forces" are being considered (and this contributes *a great deal* to their flexibility in application). Often—perhaps I should say "usually"—one characterizes a situation via specific nonfundamental forces of various sorts (e.g., contact forces, rigid-body forces) and then derives special equations, based on the constraints those forces supply, to characterize (restricted) movements in these cases.[6] The same, of course, is true of the constitutive relations of rational continuum mechanics. They provide constraints on how the material points within bodies move when subject (or not) to various forces.[7] These practices, however, no more show physical doctrine to be committed to nonfundamental forces than the application of Newtonian physics to point-masses that only exert gravitational effects on each other reveals a commitment to *gravitational forces*. The role such derived forces play in, for example, rational continuum mechanics is exactly analogous to the role fundamental forces (e.g., gravity) play in analytical mechanics: Terms for forces are ciphers, ways of characterizing more specialized laws that apply to objects in circumstances where we can't successfully apply the fundamental laws of physics.[8] It's always a substantial matter (on this view) what *form* the laws governing

---

[5] That "forces" are "additive" amounts only to it's being *relatively* easy, mathematically speaking, to characterize the movements of cohesive bodies when more than one "force" is involved: Each "force" acts—mathematically speaking—independently of the other "forces." This means that the differential equations that result from instantiating the various force laws (plus Newton's laws of motion) to various systems of cohesive bodies amount only to the *addition* of these effects (in terms of accelerations). If "forces" were not "additive," this alone wouldn't force a quantifier commitment to forces; i.e., we still wouldn't find ourselves with instantiations of the schemas in which quantification over forces occurred. *All* that would happen is that the instantiations would become *even* harder to manipulate mathematically than they already are.

[6] Two easy examples from first-year physics: sliding a block across a table, and a pendulum.

[7] See, in particular, the beginning of Truesdell's (1991, pp. 193–4) discussion of the motion of a free body.

[8] Consider the formula for a simple pendulum cited in chapter 1: $mg\sin\theta = ml(d^2\theta/dt^2)$. This is derived by transforming the "constraining forces" on the bob, among them the force(s) that prevents the bob from breaking away from the apparatus, into *geometric constraints* on the movements of that bob. This is a general phenomena: All initial references to forces dissolve into (1) geometric constraints or (2) characterizations of how velocities, positions, and accelerations are linked in how they change over time (because of those constraints), or are (3) (in part or whole) ignored altogether.

the motions of objects in an application take, but it's, at least as far as Ncm is concerned, in no sense an *ontological* matter about what sorts of *forces* are at work.[9] For talk of "forces" when applying Ncm is no more than shorthand for the forms that instantiations of law-schemas take.

This may not appear to be the case with *contact forces*, because what's more palpable than the pressure, for example, that one body exerts on another when they touch? Within the context of Ncm, however, this doesn't lead to additional ontological commitments over and above the lumps hitting each other. The constitutive relations characterize how the bodies perform under contact forces, and the latter—from this point of view—are only relations between (the surfaces of) bodies as they make contact. I say more about this shortly, but *surfaces* on which density functions are defined are part of the mathematics imposed upon lumps to facilitate the application of Ncm—they're not additional ontological commitments. As I've noted before, a more fine-tuned physical analysis (from outside Ncm) may treat contacting lumps as a very complicated process ultimately involving, among other things, the exchange of subatomic particles (or something like that), but again, all this goes quite beyond Ncm.

It may seem that this nonontological view of forces in Newtonian physics strips the latter of all hope of explaining why objects move as they do. Consider the case where the bodies in question are rigid enough, and distant enough from each other, to be treated as point-masses exerting only gravitational force upon one another. We're prone to say that such objects move as they do because of, say, gravitational *forces*—and we like to add that although there may be a mystery, in the case of gravity, about how such a force can act instantaneously over space, there is no mystery about the sources of the motions of the objects themselves: They move as they do because of the gravitational forces they exert on one another. Indeed, the picture may seem even more illuminating when we consider that objects can be affected by several forces with quite different (but additive) effects.

This last bit amounts to a pseudo-explanation, however. Different forces with different effects correspond only to the different forms the laws governing the motions of a set of bodies take under various circumstances. Although we can describe, in the gravitational case, the *cause* of a body's motions to be its mass, the mass of other bodies, and the distances

---

[9] Matters can change, of course: Scientific laws, and their applications, may be recast so that the role of terms standing for forces changes—such terms no longer operate as ciphers for mathematical constraints on the possible movements of (parts of) bodies. Further, to anticipate the way I soon approach the ontological issues here, thick epistemic access may seem to be forged to the *forces themselves*. It certainly looks like this is what's happened in the contemporary setting, because forces *are* characterized as the exchange of various subatomic particles, which we, in turn, attempt to *detect*. I regret that I must leave a more careful analysis of the place of forces in the contemporary non-Newtonian setting for another time (and place).

between those bodies, nothing is added to the explanation by mentioning the existence of *gravitational forces*.

This doesn't change when we replace the Newtonian picture of gravity with that of *general relativity*. Even though, that is (as it's commonly put), gravitational effects are part of the geometry of space-time (specifically, they're identified as *curvature* in space-time), still, the impact of that curvature is only upon how lumps move. Instantiations, that is, of the more complicated laws of *general relativity* still result in equations in which nothing that quantifies over forces remain. Of course, there is quantification over space-time points—but that's equally true of the version of Ncm under study.[10]

By the way, the claim that Ncm involves no commitments to forces doesn't mean that *causation* and *causes* are thus absent, as well, because we can describe the cause of the particular motion of one of these fairly rigid lumps *to be* (within acceptable approximation) the amount of mass within, and the location of, the lumps in the universe. Or, we can describe the lump as moving as it does *because of* the locations and masses of all the lumps in the universe. Lumps can themselves *be* causes, that is, even without gravitational forces being part of the furniture of the universe.

The last paragraph naturally raises another issue: An ontological distinction is usually made between forces and other effects that look like they're due to forces but really aren't. For example, if a forward-moving projectile hits a wall, only at that point (when it hits the wall) is a force exerted that may be great enough to make a hole.[11] That is to say, the *projectile* doesn't exert a force (*it's* only innocently moving under inertia)—the *wall* exerts a force *on* the projectile.

The distinction between inertial and genuine forces gives rise to the terminology of "pseudo-forces," of which Coriolis forces are a standard example. The *real* forces acting on a two-body system where one body rotates around (a much more massive) second body are just those forces induced by gravity (I continue to neglect the other factors involved); the appearance of a force driving the rotating body tangent to its orbit is due not to a real force, but just to inertia.[12] It may seem that this important

---

[10] "Geometrizing away forces" is, of course, already possible in the Newtonian context itself, and this can (and has) led philosophers to a more fine-grained analysis of the differences between what the geometry looks like in the (neo-)Newtonian context and its appearance in general relativity, in order to evaluate differences in how geometric commitments get into the two pictures.

[11] I draw this example from Cohen 1999, p. 99. The example is due to George Smith.

[12] It seems clear, by the way, that *Newton* was committed to the reality of forces, and despite his speaking of, e.g., "centrifugal forces" (which was meant to describe what today would be called the tangential component of curved or orbital motion; see Cohen 1999, pp. 82–4), he didn't take the latter to be real as he *did* take gravitation to be, and the other forces he considered in the course of his thinking, and this despite the similarity of language. Cohen (1999) writes: "Newton believed in and made use of many varieties of force in the course of his scientific thinking. These include—among others—'active' and 'passive' forces, chemical and other short-range or interparticle forces, and various forces associated with (or

distinction is obliterated by assimilating forces to lawful descriptions of how bodies move.

Not so: Mass, in the theoretical setting of Ncm, is a *source* of constraints on how bodies move. In particular, a sufficient (but not necessary) condition for the presence of mass is when (1) deviations of the bodies are from the motions predicted by Newton's first law of motion and (2) no other causal factors affect how bodies move (e.g., what empirically corresponds to constitutive relations, or other forces, both internal and external) that can be used to explain the deviations instead.[13] That is to say, Newton's important distinction between *vis insita* ("inherent force"), or *vis inertiae* ("force of inertia"), on the one hand, and "impressed force," on the other doesn't need to be captured ontologically by, say, distinguishing between "real" and "unreal" forces; instead, it can be captured by differences in the laws that apply to a situation and a programmatic empirical claim. The latter is this: If it seems that deviations from the first law are operating with respect to a collection of bodies, then *impressed forces* are postulated.[14] What this means isn't that an *ontological* assumption of the existence of certain forces has been made; rather, it's that a programmatic requirement of finding sources for deviations from the first law is now in place. For example, the particular distribution of mass in the universe, *or* the presence of other sources that give rise to forces, *or* bodies being composed of such and such types of materials—any of these, if they provide constraints on possible motions that prove empirically correct, satisfy this requirement. So ontological assumptions are sometimes made when impressed forces are "postulated," but it isn't that such forces are postulated to exist, but only, at best, that the sources for such forces are so postulated (and must be found).[15]

---

arising from) various concepts of aether" (p. 54). Indeed, Newton would not have been so concerned—so defensive, even—about the question of the "cause" of gravity if he hadn't believed in the existence of it.

[13] This condition is sufficient but not necessary because, e.g., the effects of mass, although present, may cancel in such a way that bodies continue to move inertially.

[14] Given that constitutive relations cause deviations from the first law, it's important for current purposes to include them under the rubric of "impressed force." Assimilating the recognition of constitutive relations to the general search for force laws may look unnatural: Constitutive relations seem much more like a *temporary* stopping point for such a program—they're to be subsequently replaced by a more fine-tuned analysis of materials that provides "genuine" force laws from which what empirically corresponds to constitutive relations can be derived. But, regardless of the success of (reductive) attempts to replace constitutive relations with "genuine" force laws, this is a fate that can befall any force law, for any of them can turn out to be temporary stopping points in the establishment of physical laws, and thus in the analysis of physical phenomena.

[15] This programmatic assumption arises again in the discussion of absolute space and time, and privileged frameworks. Notice the point of "at best"—*being made of certain materials* needn't generate ontological commitments (e.g., the existence of massive or charged objects) even though it helps satisfy the program of locating sources for deviations from the first law. See what follows.

Notice also that the program places systematic constraints on ontological assumptions:

Although gravitational forces are explicitly linked in the Newtonian context to source variables (viz., the distribution and amounts of mass in the universe), this isn't true (generally) of the forces internal to cohesive bodies or to constitutive relations, as I've already indicated. As far as Ncm is concerned, the latter are represented as entirely geometrical in character, and by this I mean both that, strictly speaking, they're taken to govern *unreal* material points, and that they codify constraints such as (1) how cohesive bodies deform (how their material points move) when forces (e.g., contact forces) act upon such bodies and (2) how forces communicate from one part of a cohesive body to another (e.g., the instantaneous communication of force from one part of a rigid-body to another).[16]

Having disparaged claims of the existence of forces, it's nonetheless extremely natural to speak of them, as my very own prose abundantly indicates. This is because of the need, when indicating the various cases implicitly characterized by law-schemas, to sweep out *over* the various forms those law-schemas take when instantiated. It's easiest to do this by invoking forces, for example, "The gravitational force exerted by a point-mass on other bodies is radial and decreases according to the inverse square of its distance from those bodies." To say this (instead) in terms of how the laws are instantiated to specific circumstances involves awkward and lengthy circumlocution.[17]

This (linguistic) factor is also behind, when describing the induction schema of Peano arithmetic (PA; see chapter 1), our falling (naturally) into talk of *properties*: If 0 has a property, and if, whenever a number has a property, its successor has that property, then every number has that property. Quantifying over the cases the induction schema applies to is facilitated if we drop awkward talk of instantiations of the induction schema and instead talk of properties of numbers. So, too, describing the cases force laws apply to is facilitated if we drop talk of instantiations of force laws and instead talk of how forces can exert effects on objects.[18]

The presence of mass, for example, affects *every* body in the universe—it won't do to postulate a massive body B to explain the deviation of some body A from Newton's first law, if the result is that B can't have gravitational impact on any other body in the universe (on pain of empirical refutation).

[16] How *forces* communicate from one part of a cohesive body to another? See what immediately follows, but also notice that what's being alluded to here is that how force *laws* are to apply to the movements of material points is constrained in various ways by the constitutive relations.

[17] Exactly the same sorts of considerations motivate the adoption of the truth idiom. See chapter 1.

[18] Newton describes his first and second laws this way: (1) "Every body perseveres in its state of being at rest or of moving uniformly straight forward, except insofar as it is compelled to change its state by forces impressed." (2) "A change in motion is proportional to the motive force impressed and takes place along the straight line in which that force is impressed." See Newton 1726, p. 416. Also, consider his proposition 1: "The forces by which the circumjovial planets [satellites of Jupiter] are continually drawn away from recti-

Although the foregoing considerations, I hope, make it clear that avoiding a commitment to forces doesn't lead to methodological complications (e.g., with the important distinction between pseudo-forces and "real" forces), it should be less clear that the substantive move of denying forces has succeeded against that opponent who, on the same sorts of linguistic grounds, thinks we *are* so committed, because resolution of the issue turns on treating Newton's laws as law-schemas and noticing that the instantiations of those law-schemas have nothing in them that (terminologically) picks out forces. But a similar argument that PA isn't committed to sets (or properties) of numbers looks like question-begging. Indeed, any instantiation of the (first-order) induction schema has no terms that pick out sets or properties of numbers—at least according to the Quinean criterion of commitment—because once the induction schema is instantiated, only predicate constants, which hold of numbers, appear, and the Quinean criterion doesn't find such predicate constants ontologically committing. But the proponent of higher-order logic—who also thinks second-order PA is committed to sets (or properties) of numbers—will claim that after instantiating a second-order variable, indications of a commitment to properties have been obliterated because they can only be recognized via, say, the (second-order) induction sentence itself: $\forall X[(X0 \,\&\, \forall x(Xx \to Xsx)) \to \forall x Xx]$. The second-order variable X in that sentence ranges over properties (or something akin to such) and thus the appearance of "$(\exists X)$" codifies a commitment to properties (or such).

Another response to the schema approach to avoiding commitment to forces is to suggest that axiomatizations of mechanics[19]—couched as they are in terms of forces—are best interpreted as involving first-order quantification over forces. On this view, when the terms for forces are instantiated with mathematical constraints of various sorts, the resulting "force-free" formulas indicate the *effects* of such forces, rather than the actual existential commitments of the physical theory.

This ontological standoff, turning on the linguistic fine print of our scientific theories, should induce déjà vu. Worse, and in any case, the linguistic approach to determining the ontological commitments of our theories seems to miss the point badly both on intrinsic grounds and in light of the discussion of ontology in part I and chapters 6 and 7 of part II.

On intrinsic grounds, the complaint goes something like this: Why should how we *massage* our theories linguistically (and in ways that seem to bear only on notational convenience) indicate what our *ontological* commitments are? The complaint isn't the mistaken one that what there is turns on *what there is*, not on what our theories *say* there is. After all,

---

linear motions and are maintained in their respective orbits are directed to the center of Jupiter and are inversely as the squares of their places from that center" (Newton 1726, p. 802).

[19] See the references in chapter 8, n. 13.

the *real* issue is what *we* think there is, and that *is* indicated by *our* theories (what we *say* there is). No, the point is that the ontological commitments of a theory should be due to how the theory links up to the world—and that's what notational reworkings of theories have (pretty much) nothing to do with.[20]

Given my foregoing discussion of ontology, the (related) objection is this: What should be relevant are the *aspects* (noun phrases, predicates, whatever) of our theories that are *connected* to what they refer to (or hold of!) in such a way that the latter must be regarded as *ontologically independent*. Linguistic massagings (and remassagings) of theories so that existential commitments are recalibrated as schematic letters (or vice versa) drop out of consideration because of their irrelevancy to methods of recognizing ontological commitment.

Instead, what needs to be shown is that there is no reason to think that *whatever* it is that's described in one notation via first- or second-order quantifiers, and in another notation via schemas, is something that's ontologically independent. And this means that no argument can be taken seriously if it presupposes that the ontological commitments of a theory must be among the quantifier commitments of a theory, because when was it shown that if something is ontologically independent then that something *will always* be represented in a first-order theory by noun phrases (and quantifiers)?[21]

Let's then reconsider the ontological status of forces from this other perspective: What reasons are there to take forces to be posits that are ontologically independent of *us*? Well, given that we take a category of somethings to exist either if we see ourselves as forging thick epistemic access to instances of it or if that category falls under a more general category that involves thick epistemic access,[22] there are lots of reasons to think we *don't* have thick epistemic access to forces: We don't (directly, or even indirectly) study *forces*. We don't determine what the individuation conditions for forces are, nor do we study what forces are *made* of.[23] And, indeed, the way that the laws of Ncm are instantiated suggests that forces are part of a mathematical formalism within which the movement of ob-

---

[20] Qualification: A notational reworking of a theory may reveal a tractable way of bringing the theory to bear on the instrumental interaction with something. But this isn't the sort of case under discussion.

[21] I *did* show (in part I) that the range of the (first-order) quantifier shouldn't be restricted to all and only what (really) exists, but when was it shown that anything that's depicted (in some way) by a theory, and that we take to be ontologically independent, must appear within the range of the quantifiers—and thus must fall within the scope of the existence predicate of that theory? Notice that I did argue in chapter 3 that *if* we want the existence predicate to hold of all and only what exists, then that predicate must be governed by a criterion for what exists. But once we've motivated a specific criterion for what exists, the possibility becomes open to us that something codified in a language in such a way that it doesn't fall under a (first-order) quantifier might nevertheless fit that criterion.

[22] See the discussion of this in chapter 6.

[23] Again, I'm speaking strictly within the Newtonian context.

jects can be predicted, but they don't provide an explanation for those movements; that is, we're not given any sense of what forces *do*, of how forces percolate from one object to another, of how they work their motional magic on objects, of how they're even *involved* in the process. All this would be perfectly reasonable as a study if forces were objects of a certain sort that we could investigate more closely to see if we could find out how they interact with other objects. Notice the point: It's not that such a study is attempted unsuccessfully, and so scientists claim they don't understand (yet) how forces do what they do. Nothing of this sort occurs in *practice*—that's what indicates the *mathematical character* of forces.

Newton (1726) describes his approach to gravitational forces (more than once) as mathematical. Here is the famous quote from the *General Scholium*:

> Thus far I have explained the phenomena of the heavens and of our sea by the force of gravity, but I have not yet assigned a cause to gravity. Indeed, this force arises from some cause that penetrates as far as the centers of the sun and planets without any diminution of its power to act, and that acts not in proportion to the quantity of the *surfaces* of the particles on which it acts (as mechanical causes are wont to do) but in proportion to the quantity of *solid* matter, and whose action is extended everywhere to immense distances, always decreasing as the squares of the distances. (p. 943)

He adds shortly after: "I have not as yet been able to deduce from phenomena the reason for these properties of gravity, and I do not feign hypotheses." And then he says: "And it is enough that gravity really exists and acts according to the laws that we have set forth and is sufficient to explain all the motions of the heavenly bodies and of our sea."[24]

Cohesive Bodies

The next issue to discuss is the claim that the reconstrual of lumps as moving sets of material points doesn't require there *really being* such sets of material points (that such are what lumps of matter really are). The first remark to make is that material points—like forces (and like spacetime points, as I discuss below)—aren't amenable to thick epistemic access. One doesn't, in order to apply Ncm, determine that lumps *really are* made up of densely packed material points: One doesn't empirically determine (via subtle instrumentation, say) that lumps are thickly com-

---

[24] For a general description of Newton's mathematical approach to forces and its application to the external world, see Cohen 1999, esp. pp. 148–55.
Also, see Truesdell (1991, p. 61), where we find: "Modern fundamental thought in mechanics has reverted to the viewpoint of NEWTON and EULER: Forces are basic, *a priori* concepts in mechanics." Translation: they're (potentially axiomatizable) mathematical concepts imposed *upon* an empirical domain. They're not interesting empirical items that we've discovered and want to truly characterize the properties of.

posed of "atoms" with no extension.²⁵ Nor does one empirically determine exactly how mass is distributed throughout lumps. (E.g., is the assumption that mass-distribution functions are continuous, integrable, etc., empirically legitimate?) Thus, the particular forms constitutive relations take aren't justified by our verifying that there are objects (e.g., material points) of certain sorts (that thick epistemic access is made to)—it's not, that is, that how constitutive relations constrain the movements of material points is verified by accessing (what empirically corresponds to) material points in lumps and seeing how *they* move.²⁶

What follows is that just as material points aren't items we forge thick epistemic access to, no more so are cohesive bodies: Cohesive bodies are designed to be mathematically tractable on the basis of their (topological) composition from material points. But the topological and geometric aspects of this composition are no more empirically verified than, say, the dimensionlessness of material points. The same is true of the surfaces of cohesive bodies and the existential commitments that accompany their mathematical definitions: Such surfaces, too, are *designed* to be smoothly amenable to certain mathematical operations; it's not that lumps are empirically inspected to see that their surfaces, and the contact forces induced, are indeed as rational continuum mechanics requires them to be (on the contrary!).²⁷ *Lumps* to which we *are* committed can't be *identified* with cohesive bodies. The ontological status of point-masses is even more obvious because these are *introduced* in applications of Ncm precisely when the internal properties of lumps can be ignored, and their masses can be treated as located at their centers of mass.

Now let's take up an issue that's been, perhaps, bothering the reader for a while: Because the laws of Ncm are couched in terms of cohesive bodies, material points, point-masses, and forces, there may be serious concerns about *what* these laws are supposed to be true *of*. Consider lumps again: Imagine first that lumps are locationally well defined in space and time; that is, that a smooth, and so on, *location function* that tells us where and when lumps occur is empirically sound. Next, assume that the distribution of the mass of any lump is empirically such that a definition of a smooth *mass function* for that lump that is parasitical on its location function is also well defined.²⁸

---

²⁵ If one *does* such a study—on items in the real world I mean—one *hardly* finds that the constituents of things are *anything like* continuum-wise distributed material points.

²⁶ One *may* empirically study how materials deform—and in doing *that*, one thickly accesses *parts* of bodies. But this doesn't bring in special commitments other than to lumps (and their macro-parts); should we go beyond that—to molecules, say—we're moving beyond the application of Ncm (*beyond*, i.e., the context of analytical mechanics and rational continuum mechanics).

²⁷ See, e.g., Truesdell 1991, p. 156, the section titled Axiom on Forces in Continuum Mechanics" and the paragraph leading up to it.

²⁸ Both of these empirical assumptions could easily turn out to be false of lumps—this is *not* to say that as approximations they couldn't be really useful for many (if not most) purposes.

Even given that the functions just mentioned in the last paragraph are empirically well defined with respect to lumps, even given, that is, that the distribution of mass in lumps is smooth and continuous (not discrete), and even given that the other defined properties of cohesive bodies apply perfectly to lumps so that the predicted behavior of cohesive bodies—how they deform and change upon impact, and so on—proves *true of lumps*, this doesn't allow an inference to the existence of the material points upon which the mathematics of cohesive bodies depends. All it allows (and shows) is that there is an *instantiation* of Ncm to a universe of cohesive bodies (which *are* composed of material points) where these cohesive bodies move the same way over time as lumps actually do.[29]

The reason to deny the "inference to the best explanation" to what empirically corresponds to material points is that despite (let's say) even a *perfect* isomorphism between the movements of lumps and the movements of cohesive bodies, we still have *no* reason to think there are (empirical) *points* (because none of this provides thick epistemic access to what empirically corresponds to material points). We *can* say that the distribution of mass in a lump is *smooth*, and we *can* capture *that* notion mathematically via *points*. But this doesn't license an ontological commitment to points. (I discuss this again below: Cohesive bodies—some of them, anyway—can proxy for something *real*: lumps. But that *doesn't require* the *rest* of the mathematical apparatus accompanying cohesive bodies, e.g., *material points*, to proxy for anything empirical—*even if that apparatus is empirically indispensable.*)[30]

That an instantiation of Ncm to a particular model of cohesive bodies gives us exactly the same results in movement as a collection of lumps careening through space means that (this particular instantiation of) Ncm is (exactly) *true of* those lumps even though the quantifier commitments of Ncm aren't the same as the quantifier commitments (in example (c)) of the empirical theory *E* of lumps. Ncm, here, is true of lumps, but it's not true of what empirically corresponds to forces, points, and so on, because there are no such things. Of course (see chapters 1 and 2), this doesn't stop the theory from being true *simpliciter*.

This conclusion raises two issues. First, one may wonder why the laws governing the movements of lumps are approached in such an ontologically indirect way (via, i.e., cohesive bodies and material points). Well, a precise characterization of lumps, and how they move and deform, which doesn't involve the axiomatizations of any terms other than those that

---

[29] Of course, I'm speaking in an ontologically noncommitting way about cohesive bodies and material points, both of which are purely mathematical entities.

[30] Recall from chapter 6 that just because we have thick epistemic access to something, it doesn't follow that we have thick epistemic access to its *parts* or to what it's *part of*, if anything. This is the companion to the observation just made about applied mathematics: The application of a mathematical system may supply a proxy role for *some* of the posits in that system (a proxy role for something we're ontologically committed to, that is). It won't, in general, do that for *all* the posits in the applied system.

refer to or hold of things we antecedently regard as existing, is most likely mathematically intractable.[31] Mathematical insight about how to prove something about something often arises from the surprising introduction of concepts that (intuitively) seem remote from the subject matter at hand. I've elsewhere (2000a, pp. 210–4) called this *implicational opacity*. Its presence is ontologically insignificant in pure mathematics because there, one cheerfully introduces new mathematical entities without fear of ontological consequences. But, in an empirical context, exactly the same activity gives rise to the apparent postulation of new entities, the *real* existence of which seems supported by empirical *laws* that require quantification over these new items (as well as, or *instead of*, the ordinary empirical entities we started with). But the necessity of quantification over something for the tractability of laws, even empirical laws, is no argument for the existence of those somethings. Once we've accepted the mathematization of empirical law, we must concede that our search for a mathematically tractable set of laws brings with it the same phenomenon (and response to it) found in pure mathematics.

And, given the situation in pure mathematics, there should be no surprises about this. After all, even if we think there really are a set of laws that govern our world, and if we mean by *that* that such a set of laws should describe (quantify over) *only* what actually exists, there is no reason to think that such a set of laws can be written down *by us*. Perhaps all that can be written down are laws that govern a great deal *more* than what's real. That, I submit, is precisely the case in Newtonian physics, and physics generally.[32]

When faced with this picture, one is naturally tempted to either deny the *truth* of the laws in question altogether, or at least to *focus on* their instantiations to what's real, and argue that only such restricted instantiations are true. Both moves are ruled out. The first is ruled out because, as part I indicated, we're committed to the truth of whatever doctrine we must apply empirically (subject, of course, to the caveats of chapter 2,

---

[31] It's hard to *prove* this, because the claim is a negative existential. But mathematical characterizations of subject matters often employ certain concepts because we (at present) have no other way to do it—and sometimes that's because there is no other (tractable) way to do it. It's also common (this should be noted) to have alternative approaches to an empirical subject matter that don't employ the same mathematical characterizations, but that employ *other* mathematical notions also going quite beyond the empirical subject matter itself.

[32] Field (1989) writes that "the most satisfying explanations are usually 'intrinsic' ones that don't invoke entities that are causally irrelevant to what is being explained" (p. 18). He contrasts these with "extrinsic explanations . . . (as when we explain the behaviour of a nonhuman or non-English speaker by reference to English sentences that he or she believes or desires)" and adds, "but it is natural to think that for any good extrinsic explanation there is an intrinsic explanation that underlies it." Natural or not, this last thought is ruled out by the considerations just raised. Limitations in our very human powers to explain things show up in the most interesting ways.

which don't apply here). The second is ruled out because an instantiation of such laws only to what we take to be real (*only*, say, to what we take to be ontologically independent) is intractable. Our remaining choice is to recognize that the laws are simply true as they stand, and simultaneously to deny that *any* of *their* existential commitments need be real.

This still allows (some of) their quantifier commitments to *proxy* for what *is* real. In case (c), cohesive bodies proxy for same-shaped lumps in which mass is distributed in the same way. Lumps are real; cohesive bodies aren't.

It's also worth adding that because Ncm (but this is true of physical theories generally) characterizes so much *more* than what's in *E* (it brings in, i.e., such a wealth of additional mathematical flora and fauna), it's far *more* useful in applications to *E*. For, as I've pointed out, we often can't apply a theory to exactly what there is because the latter is too complicated and intractable to characterize directly. Instead, idealizations are introduced, but if a physical theory has conceptual breadth beyond its capacity to describe what there is, then it can be applied without difficulty even to highly idealized contexts. In the case of Ncm and *E*, for example, the exact geometric properties of lumps and the distribution of mass throughout those lumps may be impossible to discover and characterize; nevertheless, we can still apply Ncm to all sorts of cases where we idealize the properties of lumps into something amenable to tractable physical law. A theory that described *only* what there is wouldn't be as empirically useful.

This helps illuminate the second issue raised by Ncm's failure to describe (only) what exists: Why, nevertheless, is it empirically adequate when applied to *E*? I discussed this issue in chapter 8 (elucidation (5)), and so it should already be clear that there can be several possible explanations for the empirical adequacy of these mathematical assumptions, none of which rely on the *reality* of material points, cohesive bodies, forces, and so on, and all of which turn on empirical facts about the objects of *E*. One possibility, for example, is that lumps *really are* infinitely divisible in a *smooth* way. Were this so, we could prove that Ncm gives us *exact* answers in the *limit*. But lumps could be *granular*, and then the successful application of Ncm (when coupled with mathematical assumptions about the continuous distribution of mass in lumps) would involve approximations (and in what sense these mathematical assumptions were approximate would turn on exactly how the lumps were granular).

For example (and restricting our attention to gravitation), suppose we could empirically establish that there are atoms, and that atoms are indivisible spheres (of a certain fixed diameter) within which mass is evenly distributed. In such a case, some successful applications of Ncm would turn on material points, given the tiny size of these atoms, adequately representing (at least to a certain degree of approximation) the latter.

Absolute Space and Time

The debate over absolute space and time versus relative space and time (relationists vs. substantivalists)[33] has been a seriously tangled one right from its start at the hands of Leibniz and Clarke (and Newton),[34] continuing with Mach and Einstein, then Reichenbach, and so on, until today. It's not desirable to approach this issue in anything like a comprehensive manner; apart from there being so many articles and sections of books on the subject already, my purpose is, in any case, much narrower: only to evaluate the ontological status of space and time in light of the perspective on ontology already in place.

This much, however, is worth saying to begin with. Much (if not most) of the literature on this subject presupposes that the relationist/substantivalist debate has a form similar to that of the traditional Platonist/nominalist debate over abstracta. Just as nominalists are traditionally seen as required to *expunge* mathematics of quantification over abstracta if they deny abstracta existence, so too relationists are seen as required to *expunge* physical formalism of the apparatus that refers to absolute space and are forbidden to use definitions that involve anything relationally suspect (e.g., unoccupied space-time points). Much (if not all) of the literature on this topic is dedicated to working out—with greater and greater sophistication—the implications of possible positions all of which presuppose a version of this assumption about *notation*. But, as should be clear by now, there is another way that makes the relationist/absolutist dichotomy a false one: One can use any of the physics and mathematics that substantivalists help themselves to; the question of what in that apparatus one is resultingly ontologically committed to remains open. I illustrate this moral in what follows.

It's best to start with a discussion of coordinate systems (usually called "frames"). Ncm requires a frame (a set of axes: three of space, one of time) that provide coordinates (space-time locations) for all the objects Ncm applies to. These frames differ from each other in their orientations, velocities, and accelerations. As a result, the choice of a frame has consequences. Given a point-mass in space, we can impose on that point-mass *any* sort of motion *whatsoever* by choosing the appropriate frame; the point-mass can spiral, travel in a straight line, remain motionless, and so on.

In order to (initially) leave out certain complications (forces internal to cohesive bodies, contact forces, and constitutive relations), let's first consider a universe of individual point-masses (no extended cohesive bodies). On what basis should a frame for this universe be chosen? My claim

---

[33] I should warn the reader that I continue to make free use of the term "force" and the distinction between "real forces" and "pseudo-forces." What my ontologically innocuous use of these words mean should already be clear from the foregoing.

[34] Westfall (1980) notes that it's still an open question how much Newton participated on the sly in Clarke's debate with Leibniz. Of course, there are considerations in support of absolute space and time in the *Principia* itself and in other places in Newton's writings.

is that, *in principle*, a frame should be chosen so that the point-masses move in congruence with Newton's laws of motion and his law of gravitation. That is, there can't be any accelerating point-masses (according to the frame chosen) the accelerations of which don't follow from Newton's laws of motion plus his law of gravitation. (In particular, a single point-mass in an otherwise empty universe doesn't admit of a frame in which its motion is accelerated; so, too, if a frame is chosen for a many–point-mass universe, every point-mass has accelerated motion in that universe— unless all the gravitational forces acting on a point-mass happen to cancel.)[35]

This makes the choice of a frame part of an *empirical program*: All accelerated motions arise from the distribution of *mass* in the universe and must operate in congruence with Newton's laws. The program, being empirical, can fail, of course: Perhaps *no* frame exists that satisfies this condition. Under such circumstances, the existence of other forces (with other sources) is hoped for; one then formulates the effects of these forces in laws similar to how gravitation has been characterized and tries again with the fuller package.[36]

As is well known, this program doesn't, in any case, single out a *unique* frame: If one frame does the job demanded by the empirical program, then *any* frame moving in uniform irrotational straight-line motion with respect to that first frame does it, too. That is, as it's been observed, Newton's laws are insensitive to absolute position and absolute velocity.[37] In what follows, I call the family of frames that this empirical program requires the *privileged frames*, and I call this empirical program the *privileged frame program*.[38]

---

[35] Recall that Newton (1726) shows that the sun moves because of the gravitational effects on it from the other bodies in the solar system. It just doesn't move very much.

Also, Friedman (1983) notes that, "[i]n more conventional formulations of Newton's laws of motion, by both physicists and philosophers, the notion of a reference frame—more specifically, an inertial frame—plays a crucial role. The content of Newton's First Law, for example, is taken to be the claim that free particles satisfy equation (10)[$d^2x_i/dt^2 = 0$] in inertial reference frames" (p. 116). Apart from the problem just noted (and as Friedman himself observes shortly after this quote) that there are no free particles (really), this conventional formulation gets things backward: The sources (e.g., mass) of forces come first; the reference frame (and therefore the motions, inertial and otherwise) comes second.

[36] E.g., electromagnetic forces, strong forces, etc. Of course I assume, contrary to fact, that we wouldn't be driven away from the Newtonian program altogether; empirically, i.e., the program can fail in another way (and has): The Newtonian laws of motion and his gravitational law have to be modified. Given the present aim of evaluating the ontological commitments of the Newtonian program *assuming* its continued empirical success, I'm waiving this consideration.

[37] This is often illustrated by pointing out that in a train moving with fixed velocity with respect to the ground the train tracks are on, one is—on the basis of what's felt and seen—unable to tell whether the train is moving or the *landscape* is. Although accelerations are *felt*, velocities are not.

[38] The distinction between "privileged frames" and "nonprivileged frames" corresponds to Newton's distinction between "inertial frames" and "noninertial frames."

I'll sometimes speak of *the* "privileged frame," for ease of exposition, and despite the latitude that's always available in choice of frame. Of course, it's this fact, that any of a

Expanding the domain to include extended cohesive bodies, and bringing in other factors, constitutive relations, contact forces, and so on, changes nothing essential to the empirical program. It's still required that a frame be chosen according to which all motions of cohesive bodies and their parts operate in congruence with the joint effects of body forces, contact forces, and the constitutive relations of cohesive bodies.

It should be stressed that the privileged frame program *doesn't* require that we *find* the privileged frame (for everything in space and time) in order to apply Ncm to specific cases. Here, as everywhere in physics, idealizations and measurement thresholds play an ineliminable role. We can always ignore certain effects for certain purposes.

As I mentioned earlier, the options regarding ontological commitment to absolute space and time are usually couched as a choice between two alternatives: the *substantivalist view*, which requires a commitment to absolute space (and time), and the *relationist view*, which commits its proponent to neither. And, as I've said, this is a false dichotomy: Although I deny that the privileged frame program involves an ontological commitment to space and time, it isn't that therefore the privileged frame program is a species of relationism. The latter position, at least according to versions of it I'm familiar with, requires "admissible" frames to be ones where a concrete object or a set of such objects is at rest in space (e.g., the fixed stars, as Mach suggested). This is because if velocities and accelerations aren't to be taken as genuine properties of objects but only as *relations* to other objects, then we can determine the presence of velocities and accelerations only via a frame at rest with respect to *an object* (because to take the perspective relative to an object *is* to adopt a frame according to which that object is at rest).[39] As far as the privileged frame program is concerned, however, the privileged frame needn't (and generally won't) be one in which there are *any* nonaccelerated objects. That this (generally) implies the privileged frame is "anchored in" no particular concrete object may seem to impel a commitment to "absolute space" because the coordinates laid off in this chosen frame are thus relative to points and times in sheer space (and time) themselves. Even if this result requires quantifying over places and times where no object is located, on

---

family of frames, as I've characterized them, will do the job, and not a single one alone, that shows that, mathematically speaking, what's *indispensable* to the application of Ncm is not the Cartesian product $\mathscr{R}^3 \times \mathscr{R}$, with the standard metric, but a much weaker structure, a four-dimensional real affine space. This particular wrinkle isn't of concern in what follows because, in any case, indispensability is insufficient for ontological commitment. On the mathematical situation regarding Newton's theory, see Stein 1967 or Friedman 1983.

[39] This isn't the *whole* story about possible versions of the relationist view. See Friedman 1983, esp. pp. 223–36. Also see Maudlin 1993. Because my only point is that the privileged frame program *isn't* a relationist view—quantification over *all* the apparatuses that substantivalists help themselves to is acceptable—there's no need to sort through this particular tangle.

the view of ontological commitment developed in part I and chapters 6 and 7, this (alone) won't make such posits other than ultrathin.

It's important to keep in mind the separation (so crucial to the argument in part I) of what's *true* from what *exists*. Both the *objectivity* of the privileged frame being the *right* one and, therefore, the *objectivity* of the claim that objects are indeed in the grip of accelerated motions (or not), because they're seen to be so from that perspective, are preserved by the privileged frame program. However (I repeat), this objectivity (alone) *doesn't* require an ontological commitment to space or time (or spacetime points), even if it *does* require spatial and temporal posits—and even if, I must add, the posits required (and indispensable) are "absolute ones."[40]

What about the objectivity of the question of whether objects are at rest or moving? This issue proves to be empirically indeterminate, because privileged frames aren't distinguished by differences in their linear velocity. But, again, *that's* not a problem unless we feel that there is some fact about the objects—their relation to space and time—that we're missing out on (and to even *worry* about this presupposes a substantivalist view about space).[41]

I eventually consider the counterargument that the role of space and time in *Newtonian physics* shows they *can't* be ultrathin—that is, that they *must* be ontologically independent of us and, in addition, must be items to which we (sometimes) have thick epistemic access. But before taking this issue up, I want to note that Stein is especially scornful of certain views of Alexander's and Toulmin's and expresses this scorn in a way that suggests it can be applied to the view on offer.[42] Stein writes (1967) of passages he has just quoted from Alexander and Toulmin that

---

[40] Stein (1967) notes that Newton applies the adjective "absolute" "in its logical sense to concepts that are not 'relative' to something" (p. 189). He elucidates this claim in a footnote:

> For instance, what Newton calls "absolute quantity" of the earth's gravity is a property of the earth itself, which we call the earth's "gravitational mass"; in contrast with the "accelerative quantity" of the earth's gravity, which varies from place to place and is a property of a point in a given position relative to the earth—the intensity of the earth's gravitational field at that point; and in contrast with the "motive quantity" of the earth's gravity, which depends both upon the place and the body acted on—the weight, or attraction exerted upon the body by the earth. (p. 199)

This is the sense of "absolute" as I use it.

[41] Does this mean I'm giving up on bivalence? It may seem that ontological issues about space and time translate into ones about the assignment of truth values to sentences. They don't. If one is very prissy about the matter, one can reformulate the theory to minimize such empirically indeterminate truth values (this is one way to see the move from the traditional geometry for Newtonian physics to neo-Newtonian space-time). But another option is to *not bother*. We can adopt bivalence and take ourselves "not to know" certain truths. See my 2000b, pt. IV, § 6, where I discuss this take on the epistemic idioms further.

[42] Qualification: I'm *not* saying that the views Alexander and Toulmin express *are* mine. Alexander says, e.g., in a remark disapproving of Newton, that "Newton thought he was doing more than just identifying the set of frames of reference with respect to which

[i]n both of these passages one observes what I should characterize as a loose and uncontrolled manipulation of verbal distinctions. What exactly do these authors mean by "ideal entities which it is helpful to consider in theory," or by a notion or theory that "has a physical application," —as opposed to entities that "exist in reality." ... If the distinction between inertial frames of reference and those which are not inertial is a distinction that has a real application to the world; that is, if the [affine] structure [of space-time] is in some sense really exhibited by the world of events; and if this structure can legitimately be regarded as an explication of Newton's "absolute space and time"; then the question whether, in addition to characterizing the world in just the indicated sense, this structure of space-time also "really exists," surely *seems* to be supererogatory. (p. 193)

And he adds a few lines later: "It is important to be very clear about one point: the notion of the structure of space-time cannot, in so far as it is truly applicable to the physical world, be regarded as a mere conceptual tool to be used from time to time as convenience dictates. For there is only one physical world; and if it has the postulated structure, that structure is—by hypothesis—there, once for all."

However, even though the notion of space-time is *not* a mere conceptual tool to be used from time to time as convenience dictates, it does *not* follow that the structure is *"there"*: This claim presupposes an indispensability thesis—if not the *Quine-Putnam* indispensability thesis, then some indispensability thesis (or other) painfully close in import to that one. Contrary to such a thesis, indispensable structure in an empirical theory, and the posits that accompany it, needn't exist.[43]

I now make several observations in order to take up various concerns about the privileged frame program that the reader may have: First, I should point out that I'm assuming that stable units of time and space and stable ways of measuring distances and durations are available. This assumption *doesn't* require the existence of "absolute" space and time, although, of course, it *does* require the positing of space and time (and the points within them), as well as those posits of space and time having certain properties (isotrophy, etc.). But it's clear by now, I hope, that the mere presence of a posit is insufficient for ontological commitment to it, and measurement practices, turning only on the *mutual consistency* of the results of measurement *alone*, seem to be a perfect example of a body of practices that *don't* have as their goal the squaring of the claims we make

---

the laws of dynamics would take the simplest form" (cited in Stein 1967, p. 192). That the goal of the privileged frame program is to identify the set of frames with respect to which the laws of dynamics take the simplest form is *not* my view: The aim is to find a frame so that no motions of objects or their parts occur that aren't described by Newton's laws of motions, the force laws, and the constitutive relations.

[43] Having said this, it's nevertheless appropriate to add that Stein's remarks are best seen as a *challenge*: Such language as "fully exists" and "exists" (or "really exists" and "exists," as I'd prefer the contrast to go), if distinguished, must be distinguished in a *principled* way. See part I, esp. chapter 5.

with something independent of us (i.e., the properties of *actual* space and time). *This* point about our measurement practices—that they seem to have no connection to space and time, if the latter are understood to be ontologically independent of us—has been made *against* the Newtonian assumption of absolute space and time, and indeed, against earlier absolutist views of space and time, more than once.

Second, there may be a fear that the privileged frame program is incoherent because it circularly requires information (e.g., of mass values) that it presupposes possession of. I now argue that the program isn't ill-defined and that impressions to the contrary turn on (sometimes implicit) verificationist assumptions.

Imagine, to begin with, two point-masses in space, the objection may begin. *Given* the amount of mass possessed by each point-mass, the effects they have on each other are determined, and therefore the privileged frame (*modulo* differences in velocity) that their motions should be characterized from the perspective of is determined, as well.

But, the objection continues, this immediately raises a worry about the supposed sensitivity of the privileged frame program to empirical data. Put it this way: *The privileged frame program apparently presupposes that there is a way of determining the mass of something independently of Newton's laws.* This is because only if the masses of the objects in the universe are fixed does it follow that their accelerations are fixed. But if we determine the mass of something via Newton's laws, then it seems possible that adopting a different privileged frame (in some cases) can be squared with Newton's laws (so that all motions of the objects accord with those laws) by simply adjusting the assumptions about masses that the objects are taken to have.

This circularity objection turns on an (implicit) confusion of epistemic with ontological issues. The privileged frame program *does* assume that the amount of mass in the universe (and the relative distances of instances of such massive objects from each other) is given to begin with—that's an *ontological assumption* (built into the program by the relationship postulated between accelerations—due to gravitation—and mass). Ncm is committed to the existence of mass, *however* its presence is recognized; the point, then, is that given a distribution of mass in the universe, it follows from Newton's laws that a privileged frame is determined, regardless of whether *we* can determine what frame that is. It's *not* assumed, however, that facts about mass can be "determined" (i.e., discovered by us) in a way that's independent of our use of Newton's laws.[44] However, in most circumstances, the privileged frame is empiri-

[44] But *must* methods of recognizing mass presuppose Newton's laws? Of course not: "Mass" is an empirical term, not a term with a fixed definition. This means that "mass," like any other empirical term, is always open to new ways of gaining access to what it refers to, and what properties that has, and this point holds even if the Newtonian laws aren't superseded. See my 2000b, pts. III and IV, for details on empirical terms that are, as I call them, *criterion transcendent*.

cally accessible: Relative *differences* in mass can be recognized experimentally via frames that aren't privileged because, for example, the accelerations of two objects (more or less isolated from the gravitational effects of other objects) toward each other (when one or the other one is fixed) won't be the same unless they have the same mass.[45] By these sorts of experiments (and once a unit of mass is in place), we can (in principle) assign mass values to every object in the universe and so determine the privileged frame.

The *ontological assumption* that the distribution of mass throughout the universe is independent of us and our attempts to determine it extends to the other sources of forces in Newtonian universes, and to constitutive relations, as well. Regarding constitutive relations, the properties of materials, their responses to contact forces, and so on, are also matters independent of our capacity to determine them, and this means that how forces affect those materials are also ontological givens. And, as I showed in the case of mass, all of these factors are taken to determine a privileged frame wherein all deviations from Newton's first law are due only to such factors, and, of course, none of these factors prove idle—for example, no object travels unaffected by the mass of other objects without compensating forces (or constitutive relations) that explain this.

These ontological assumptions, however, don't imply that the privileged frame can (epistemically) always be discovered: As I show later, we can imagine universes that obey Newton's laws but in which it's empirically impossible to determine the privileged frame, even given access to Newton's laws, all the relevant force laws, and all the relevant constitutive relations. But it's possible to determine the privileged frame (within measurement thresholds) in *our* universe (as far as we can tell and, again, presuming the correctness of Newton's laws), and unless one *really is* a verificationist (and denies that a fact of the matter *is* a fact of the matter *unless* there is a method for determining what that fact is regardless of epistemic circumstances), that there are universes in which the privileged frame can't be empirically determined isn't relevant to the cogency of the privileged frame program.

Imagine this case:[46] a strangely empty universe except for (i) a circular disk (with uniform mass) spinning in space (around an axis through the center of the face of the disk), and (ii) a torque (never mind *what sort*,

---

[45] The more massive object accelerates more slowly toward the less massive object (when the latter is fixed) than when the situation is the other way around. This is a quite crude description of an experiment of this sort. Others (many others) much more nicely designed are possible (and actual). See Stein 1967, pp. 190–1, for a good, although brief, discussion on this.

[46] The reader will have noticed that I often try to carefully distinguish *terminologically* the *mathematical* items, cohesive bodies, forces, and constitutive relations, from the *empirical* items, lumps, sources of forces, and the materials out of which lumps are composed. I've perhaps not been perfectly consistent in this respect, but I'll be even less so in what follows, for the sake of exposition.

exactly) applied to its side that continuously increases its rotation. Eventually, if the disk rotates fast enough, it will tear apart as its subbodies move tangentially to the orbit the constitutive relations (or internal forces) hold them in. Notice that the only factors relevant to this example are (1) the contact force applied to the disk (and transmitted to its subparts), (2) the constitutive relations (or internal forces) serving the twofold role of transmitting the torque through the disk, and holding its material points in place, and (3) the (negligible) gravitational forces induced by the presumed uniform distribution of mass in the disk.

If the constitutive relations are *finitely* powerful, and if their impact on the disk's cohesion with respect to inertia is measurable—which, empirically, is to be expected—then one can (empirically) determine how fast the disk is rotating because a privileged frame is fixed by when the constitutive relations allow the disk to become distorted. If, however, the constitutive relations of the disk make it a rigid body impervious to distortion, then no matter how long the torque is applied to the coin, there is *no way* to determine its rotational velocity empirically because there is no way to determine what the constitutive relations are actually doing. In such a case, we can't determine (empirically) a privileged frame because the infinitely powerful constitutive relations of the disk make it impervious to empirical study (given the restricted resources in this universe). If, however, there were disks of other materials with internal forces or constitutive relations of finite strength, the privileged frame could be determined from *them*, and then by imposing that frame on the original disk, the work being done by *its* constitutive relations could be determined, as well. In practice, because what materials that objects are made of affects how they respond to forces in a systematic way, and because the privileged frame applies to *all* of space and time, the study of *any* spinning shape is epistemically legitimate in our search for the privileged frame. In a universe strangely devoid of objects, such practices can't be carried out, but there is nothing wrong with the (ontological) assumption that the privileged frame is given nonetheless by how the constitutive relations (of a disk in space, say) *actually are* causing deviations from Newton's first law, and even if epistemic access to them is ruled out.[47]

Third observation: Imagine that it turned out that (certain) frames differing with respect to their *accelerations* were compatible with the same (relative) velocities, accelerations, distances, and so on, of (particular configurations of) a set of bodies. Would this create problems with the frame program? It's hard to see why. It *would* show that empirical results about the distribution of mass would *underdetermine* our choice of frames more than it appeared at first sight. But so what? If, that is, frames *really are*

---

[47] Am I shoring up my realism about the facts of accelerations in such a universe via a kind of modal realism—by access to facts about what *would be* the case if there were other objects in such (ontologically deprived) universes? I think not, although a discussion of this takes us beyond the scope of this book.

impositions of a mathematical framework, why is this a particularly big deal? Why, that is, would this imply that *space* is something in and of itself? One imagines, I suspect, that *something else* is (empirically) determining which set of accelerated frames are actually instantiated in our universe, and then one leaps to the claim that that something is *space*. But this really is an ontological leap unless we've got independent access to this "space." Otherwise, one has to say here what one has to say in other cases where specific branches of mathematics are applied, and we can't see why: This model of the mathematics *works*, that one *doesn't*, and *we don't know why*.[48] As I've said above, this issue doesn't arise as far as Ncm is concerned, but it does arise in general relativity. This message still applies.

Fourth observation: I claim that in no way should the privileged frame *itself*, or the posits it refers to, be considered *causally* relevant to the accelerated movements of objects (from the perspective of that frame), because it's clear from the empirical frame program that such accelerations are to be explained entirely by Newton's laws of motion, the force laws (i.e., the *sources* of forces), and the constitutive relations. Nothing remains for the frame itself, or the absolute space and time it purportedly picks out, to *do*.[49] I spend the rest of the discussion of case (c) on this last observation.

Let's start by recalling two famous experiments, which Newton describes and which a number of philosophers and physicists have taken as falsifying the claim just made.[50] Although Cohen (1999) notes that these experiments are "often wrongly believed to provide the basis of his argument [for absolute space] rather than to serve as an illustration of it" (p. 107), I briefly describe these experiments and argue that they don't even *illustrate* that claim.

A bucket of water hangs from a twisted rope so that when the bucket is released, the untwisting rope spins the bucket. At first the water in the bucket is motionless relative to the ground. Eventually, the water starts to spin relative to the ground, until it's spinning so that it's motionless relative to the sides of the spinning bucket. Finally, when the bucket is

---

[48] The reader may still feel that what the mathematical term "space" picks out in this application—viz., physical space—is the culprit. I discuss why we should resist this take on the mathematical formalism toward the end of the discussion of the fourth observation.

[49] If right, this, coupled with the first observation about how stable measurements of durations and distances are found, forecloses on the possibility of forging thick epistemic access to space and time and thus leaves us with no reason to regard the posits of space and time as anything more than ultrathin.

[50] Some philosophers have taken these experiments as showing that the Newtonian view attributes dynamical properties to space itself. Others (e.g., Mach) have seen them as a challenge: to provide a frame, anchored in some concrete object or other, relative to which the accelerative effects depicted can be seen to occur; and they have denied that such accelerative effects will occur if such concrete objects (e.g., fixed stars) are missing. Still others (e.g., Einstein) have used a variant of these thought experiments to run a verificationist objection to absolute space and time as they occur in Newton's system.

stopped from spinning, the water continues to rotate, and further, the rotating water recedes from the center of the bucket and up its sides. The second (thought) experiment is of a universe with only two spheres connected by a rope. If they're rotating with respect to each other, the rope becomes taut as they try to tangentially move from the orbit the rope holds them to. If they aren't so rotating, then the rope remains slack.

Do these examples show that space *itself* has causal properties? If one accepts the privileged frame program, it's hard to see why. Consider the two spheres first: There are internal forces (or, if you wish, constitutive relations) *in the rope* that can hold the spheres in orbit. Thus, if the rope become taut, the accelerated motions involved (the spheres *continuing* to circle around the center of mass of the system instead of flying tangentially away from each other) are due solely to the factors within that rope. Thus, the choice of frame is determined as one in which the spheres are rotating rather than one in which they're motionless. More can be said about this example: The spheres themselves are composed of subbodies held, let's say, rigidly together. Were these internal forces or constitutive relations absent or weak, it's not just that the spheres themselves would fly away from each other tangentially to the orbit they're bound to by the rope, but that they would fragment into smaller bodies that would individually fly off tangentally.[51] Recognition of these accelerated motions (of the subbodies within the spheres) is satisfied in the privileged frame by their source in constitutive relations or the sources of other (internal) forces. On the other hand, if the rope remains slack, it's clear that there are no accelerated motions involved in the example at all, and a frame in accord with this must be chosen instead.

The analysis of the bucket case is almost the same, except that in this case the relevant constitutive relations and contact forces occur not in a rope (and within the spheres) but within the fluid itself, and between the fluid and the bucket, because it's these factors, and the obvious conclusion that it can only be them reacting against inertia that causes the fluid to recede from the center of the bucket and up the sides of that bucket, that force our choice of a privileged frame in which, at first, the bucket is spinning (and the water isn't) and then, later, the water in the bucket continues to spin (although the bucket no longer is).[52]

A similar lesson can be drawn from an example that Einstein (1916) gives:

> In classical mechanics ... there is an inherent epistemological defect which was, perhaps for the first time, clearly pointed out by Ernest Mach. We will

---

[51] This depends, of course, on exactly how the factors internal to the spheres operate and how the internal factors in the rope transmit forces to the subbodies in the spheres.

[52] Given the global nature of the frame to be chosen, however, an analysis of the internal forces and constitutive relations of the *twisted* rope is relevant and enables us to fix upon a frame in which, subsequently, the bucket spins, and later, the water does also. For that matter, as I've mentioned, fixing the frame via something else (far away) will do the job too!

elucidate it by the following example:—Two fluid bodies of the same size and nature hover freely in space at so great a distance from each other and from all other masses that only those gravitational forces need be taken into account which arise from the interaction of different parts of the same body. Let the distance between the two bodies be invariable, and in neither of the bodies let there be any relative movements of the parts with respect to one another. But let either mass, as judged by an observer at rest relatively to the other mass, rotate with constant angular velocity about the line joining the masses. This is a verifiable relative motion of the two bodies. Now let us imagine that each of the bodies has been surveyed by means of measuring instruments at rest relatively to itself, and let the surface of S_1 prove to be a sphere and that of S_2 an ellipsoid of revolution. Thereupon we put the question—What is the reason for this difference in the two bodies? No answer can be admitted as epistemologically satisfactory, unless the reason given is an *observable fact of experience*. The law of causality has not the significance of a statement as to the world of experience, except when *observable facts* ultimately appear as causes and effects.

Newtonian mechanics does not give a satisfactory answer to this question. It pronounces as follows:—The laws of mechanics apply to the space R_1, in respect to which the body S_1 is at rest, but not to the space R_2, in respect to which the body S_2 is at rest. But the privileged space R_1 of Galileo thus introduced is merely a *factitious* cause, and not a thing that can be observed. It is therefore clear that Newton's mechanics does not really satisfy the requirement of causality in the case under consideration, but only apparently does so, since it makes the factitious cause R_1 responsible for the observable difference in the bodies S_1 and S_2. (pp. 112–3)

This example is neat because, as Einstein describes it, we're to treat the internal forces as ones that allow the particles to respond motionally in a continuous manner when affected by other forces or by inertia. The internal forces in these bodies, that is, aren't infinitely powerful or ones that allow bodies to suddenly respond to other forces only when certain thresholds have been reached.

I gloss (somewhat diffidently) Einstein's reasoning this way: First, the differences in the two fluid bodies S_1 and S_2 are taken, on the Newtonian picture, to be due (causally) to the fact that S_1 is rotating with respect to absolute space and that S_2 isn't; or, better, that S_1 is rotating relative to the privileged frame R_1 and S_2 isn't. But surely a *frame* is a factitious cause; if someone objects that it isn't the frame that causes differences in shape between S_1 and S_2, but differences in rotation with respect to absolute space itself, the response is that such an answer isn't epistemologically satisfactory because rotation with respect to absolute space isn't an "*observable fact of experience*."

The contribution the proponent of the privileged frame program makes to this debate is that, in any case, the *frame* isn't a causal agent, nor is the space represented by such a frame. The causal agents are *only* the sources of force that, in this case, give rise to the forces internal to the bodies. Given that these forces are (not so) rigid, it can be claimed that the point-masses[53] of S_1 aren't (attempting to) move inertially against such forces because otherwise S_1 would change in shape. This fixes what (set of) frame(s) are appropriate for describing the motions of S_1. In the case of S_2, clearly, its point-masses *are* moving (to some extent) inertially against the forces holding S_2 together. This fixes the (set of) frame(s) appropriate for S_2.[54] The only causal factors involved in the example are the gravitational forces acting between the smaller bodies within S_1 and S_2, respectively, and the vanishingly small effects of gravity between the fluid bodies S_1 and S_2 themselves.

I should say, at this juncture, that Leibniz seems to have something like the frame program in mind. He writes: "I grant that there is a difference between an absolute true motion of a body, and a mere relative change of its situation with respect to another body. For when the immediate cause of the change is in the body, that body is truly in motion; and then the situation of other bodies, with respect to it, will be changed consequently, though the cause of that change be not in them" (quoted in Friedman 1983, p. 228).[55]

Friedman (1983) dismisses the suggestion curtly: "[A]s we have emphasized many times already, uniform rotation is *not* correlated with external forces in Newtonian theory" (p. 225).[56] This is true but seriously misleading because it gives the impression that a solid sphere, for example, rotating in empty space, involves no "impressed forces." (Angular velocity is conserved, and the object continues to rotate forever, if no other forces intercede.) But this is a mistake. Forces (and here, in accord with my terminological convention of chapter 8, n. 12, I'm *including* constitutive relations) *are* required if the sphere is to continue to rotate *as a sphere*, but of course, these are forces *internal* to the object; nevertheless, these forces *are* external (or "impressed" forces, in Newton's terminology) to the smaller bodies within such spheres, because it's such forces that ensure that the object *remains* a sphere and doesn't fragment apart

---

[53] I'm assuming that the example presumes fluid bodies to be made up of (quite small) particles that exert nonzero gravitational forces on each other rather than continuous bodies with continuum-many mass points with zero mass but with a nonzero mass function defined over each body. The lessons I want to draw are available in either case, but I'm trying to stick as closely as possible to my reading of Einstein's text.

[54] And these turn out to be the *same* family of frames. It's, of course, a *very important* empirical assumption of the privileged frame program that the same family of privileged frames applies to *every object* in the universe: a failure in this is taken (initially) to imply that other (unaccounted for) sources of force or other factors are locally present. A more radical proposal would be the denial that a single family of frames does apply to everything, but this requires deserting the Newtonian context.

[55] It's clear that Leibniz is focused on contact forces, and Friedman rightly takes him to task for this (there's no avoiding action at a distance in Ncm). But even if we drop *that* constraint, Friedman thinks the program still fails to handle accelerations arising when constant rotational velocity is present; this is wrong, as I show in a moment.

[56] Dainton 2001, p. 177, repeats the dismissal as curtly.

into bodies traveling tangentially to the orbit they're (individually) trapped in.[57]

One factor that pushes one's contemplation of these examples in the direction of the thought that space itself must be a causal factor relevant to whether bodies or their parts are experiencing accelerations is the global nature of the constraints of the privileged frame program. This is because, according to the privileged frame program, how objects move in *any* part of the universe constrains what frame can be chosen—and therefore what such objects do (and what they're made of, etc.) affects what accelerations can be attributed to any *other* body, regardless of its distance from everything else. But this global constraint of the choice of a privileged frame is *not* glossed as a causal factor affecting bodies (although this is tempting): It's a mathematical coding of the details of the privileged frame program—that no accelerations are present save those due to admissible factors: mass, charge, constitutive structure, and so on.

Despite this, some may still be tempted, no doubt, to think that the successful application of such an apparatus as the privileged frame program (and its accompanying Euclidean geometry), even if described as purely mathematical, must be due to something—a truthmaker (or, more accurately, an *application-maker*)—in which properties inhere and to which the correctness of the application of these mathematical constraints is clearly beholden, that is, space itself. But this, I suggest, is a thinly disguised desire to provide an explanation for the success of (part of) a theory at a stage where no *genuine* explanation is ontologically available.

---

[57] Dainton (2001) writes: "A rotating solid disc will continue to rotate at a constant angular velocity, without any need for any external force or torque to be applied: it is maintained in motion by the preservation of angular momentum. . . . And yet, if the disc rotates fast enough, the centrifugal forces [inertial forces!] generated by its rotation will tear it apart" (p. 177). The last sentence is *nearly* on the point (speaking of inertial forces "generated" by rotation is, I think, misleading), but one must focus not on the disk being "torn apart" but it's *remaining intact* (at any rotational rate, however low): That's where *forces* come into play. (Recall the remark about the wall in the paragraph to which n. 11 is appended.)

Sklar (1976) makes a similar objection to the suggestion I'm attributing to Leibniz:

[O]ne seems to be able to imagine situations where there are inertial forces but no causal forces inducing the motion at all. Again imagine a uniformly rotating universe of material objects. Now imagine it to have been always in rotation, the rotation not caused by any imposed torque at all, but simply eternally sustained by the conservation of angular momentum. There would still be inertial forces present on any object at rest in this coordinate frame. (p. 192)

Nothing (that's extended), whether a disk or a universe, can rotate (from eternity or otherwise) unless there are constitutive relations or other *internal* forces holding it together.

I suspect that what's involved here is an overreaction to the fact that angular momentum, like linear momentum, is conserved (without an external torque, angular momentum remains constant). But the issue, of course, is whether *nonzero* angular momentum is possible without what amounts to *internal* forces. The answer is no, but it is yes in the case of linear momentum. This is why "inertial forces" arise in the one case but not in the other.

We can always fatuously postulate a something or other stipulated to have the properties that make it amenable to, say, the privileged frame program and Euclidean geometry. It's hard to see why such sheer postulation should be seen as *explaining* anything. Why not just admit that the global constraints of the mathematics of the privileged frame program *are true*, and (at least at this stage) we don't know why.

Because the imagery of space, time, and space-time has such a powerful grip on our intuitions, the considerations of the last few paragraphs are worth expanding on. I start with the intuitively powerful suggestion that space is not *nothing*. Newton writes (in *De Gravitatione*—I take the quote from Dainton 2001):

> [Space] is not an accident. And much less may it be said to be nothing, since it is rather something, than an accident, and approaches more nearly to the nature of substance. There is no idea of nothing, nor has nothing any properties, but we have an exceptionally clear idea of extension, abstracting the dispositions and properties of a body so that there remains only the uniform and unlimited stretching out of space in length, breadth and depth. . . . In all directions, space can be distinguished into parts whose common limits we usually call surfaces; and these surfaces can be distinguished in all directions into parts whose common limits we usually call lines. . . . Furthermore spaces are everywhere contiguous to spaces, and extension is everywhere placed next to extension. . . . (p. 153)

And Dainton (2001) picturesquely describes space as "the unseen constrainer" and asks, if space is nothing,

> why should our possibilities for movement be so tightly constrained? If space is pure void, not only would we be free to wander in the additional directions made available by a fourth dimension, but there would be nothing to prevent our wandering in the limitless different directions made available by 5-, 6-, 12-, 101-and *n*-dimensional spaces, which are just as possible, mathematically speaking, as 1-, 2-, and 3-dimensional spaces. (pp. 134–5)

Both Newton and Dainton, in these respective quotes, contrast the substantivalist view of space with the void conception. Dainton (2001) also writes later: "[S]ince a pure void has no determinate structure or dimensionality, the void-conception of space leaves something unexplained—why it is that our movements are constrained in the ways they are" (p. 224).

The emptiness of the explanation offered by the postulation of an entity—space—with the properties needed is masked by posing a false dilemma between the substantivalist view and the void conception. The right alternative to the postulation of absolute space isn't the claim that "space" is *nothing*; the right alternative is to note that "space"—strictly speaking—is a mathematical posit of the formalism of physics applied to the world, and that therefore its empirical adequacy (as part of this entire theory) is something we currently have no explanation for. *One* possible explanation, of course, is that the mathematical posits in a successfully applied theory *exactly correspond* to the physical world (there are, in the world, objects with exactly the properties the mathematical entities are postulated to have). But, apart from the global success of the applied mathematics, we've no reason to think this is true, and because there are *lots* of reasons for why an empirically adequate theory could be empirically adequate (as the history of science indicates) *apart from* the entities that theory postulates actually existing, and because, in any case, without thick epistemic access to those posits, the theory is a black box, we've no epistemic justification for this easy solution.

When it comes to fiction, we can all agree that the properties attributed to fictional objects are *in no way* beholden to things independent of us. But as the foregoing observations make palpably clear, such isn't true of space: The properties attributed to it are ones we have no choice about because they impose on how objects move—and that's not a matter for us to dictate. The movements of objects through space, that is, are themselves in thrall to something ontologically independent of us and, indeed, ontologically independent *tout court*.

This much is true, but we can't say more about what this something is, and just because, in imposing a branch of mathematics, entities are brought along that we attribute specific properties to, and just because we've no alternatives in what properties we can attribute to such objects (and still have the applied mathematics *work*), and despite our inability, say, to imagine lots of alternatives, it doesn't follow that we've managed to get *our terms* ("space," say) to refer to that something that is ontologically independent of us and that is the source of the indispensability of this application of pure mathematics. We *may* have, but in lieu of specific access to the items referred to by terms in the mathematical formalism, we've only got the global success of that applied mathematics as an argument for the claim that such terms refer to what's ontologically independent of us, and as I've already shown, that won't suffice.

As an illustration of how the spatial posits are operating in a purely mathematical way, consider again the Newtonian story about rotating disks. Mach once suggested that inertial forces are produced when a body accelerates with respect to the center of mass of the entire universe. Dainton (2001, p. 179) suggests that this requires a mechanism by which the center of mass can do such a thing. It's an implication of the frame program that, restricting our attention to gravitation, inertial effects can be observed if a body accelerates with respect to the center of mass of the entire universe, but the need for a *mechanism* is denied because it isn't that the center of mass is causing anything: It's that, when all the accelerative effects of the (real) forces have been accounted for, this result pops out as a corollary.[58]

---

[58] But why does it pop out as a corollary, one may wonder; why should this global version of the frame program work? That's to ask for an explanation where no explanation (from within Ncm, that is) is possible. And this is always the case: We can't reach a point

But now consider *this* explanation: Space is everywhere and everywhen, and so the *mechanism* by which rotating disks and Newton's spinning bucket experience "inertial forces" is by their parts acceleratively "rubbing up against" sheer space (and time) or, less derisively, accelerating with respect to the points of absolute space and time (in their *local environment*). This explanation is a nonstarter, not because it couldn't be true, and not because it couldn't be that we could achieve thick epistemic access to spatial points to see how they manage to do this to the objects that move through them, but because that's not how the theory *works*: The mathematics of space isn't an empirical theory of how inertial effects arise in rotating objects; it's the postulation of laws of motion with respect to mathematical posits (and the brute truths about the motions of objects in Ncm with respect to space and time don't—*as they stand*—allow us to draw any ontological conclusions). The whole thing works—but we have no access to any explanation why or to what sorts of objects are involved.[59]

The last point I want to make about this addresses the possibility of thick epistemic access to absolute space. Nerlich (1994a, p. 38; 1994b, § 7) gives the example of the non-Euclidean hole. This is a case of local curvature in space (the size of a football, say) that is so drastically different from the surrounding space that it can be seen (because of how light bends within it) and its contours determined. Furthermore, such a space (in principle) could affect the other senses, as well. Given such a thing, wouldn't we say that we had thick epistemic access (indeed, *observational* access) to space itself?[60]

The answer is no. Consider an ordinary table: I have observational access to it, and in particular, I can press my hand on its surface. As it turns out, the actual (subatomic) matter making up such a table is very rarely distributed throughout that table. What prevents my hand passing through the table are the electromagnetic forces generated by the electron constituents of the table. Do I have thick epistemic access to these

---

where the successful application of *all* the mathematics in a theory is explicable (because a fundamental theory always employs mathematics indispensably—and we can't explain the success of that mathematics from within the fundamental theory itself). My point at the moment is not so much to illustrate this simple fact of scientific life that the Quine-Putnam indispensability thesis starts from, but rather to show how we sometimes attempt pseudo-explanations by (what amounts to) postulating physical entities that mimic the properties of the ultrathin posits of the applied mathematics.

[59] Contrast this with the forces that affect objects moving through a fluid such as water. Here we have access (in principle) to the mechanisms to which such forces are due—how the molecules of water move around (and adhere to) moving objects—and what sorts of (molecular) forces there are. Notice also that we've got access—thick access—to the objects in question: not just to the objects moving through the water, but to the components of the water itself. This blatantly contrasts with absolute space and time in the context of Ncm.

[60] Nerlich (1994a) writes: "[W]ould we say, not just that we see *that* there is such a hole, but that we *see the hole itself*, just as we see warps in clear glass?" (p. 39). Also see Dainton's (2001) discussion of this.

electrons and to electromagnetic forces? Not by virtue of my hand I don't. The theory of the table (which I've just alluded to) describes it as composed of molecules doing such and such, and a number of the items in that theory are items to which we've forged thick epistemic access. Forces may be among these—but none of this indicates that what I *see*, when I look at a table, are such molecules or that what my hand is pressing against are such molecules (it's pressing against *a table*).[61]

In the same way, we may find that the best physical theory for such observable holes is one that describes them as curvature in space. But it's an additional question whether *space* as such exists: This turns on thick epistemic access not to the (macro)holes but to the purported constituents of such holes (points, say). Under such circumstances, I would say that we still would not be committed to space, or to its curvature; we would simply have a true theory (treating such "holes" as local curvatures in space) that proved to be globally accurate but that told us nothing about the actual constituents of such holes and that, within the context of the theory, allowed no instrumental access to such constituents.

Space-Time Points

I now turn to a discussion of space-time points themselves (largely apart from, i.e., the question of the existence of space and time themselves).[62] Field (1989) has argued that the reasons for refusing a commitment to *abstracta* aren't available for the rejection of "space-time regions," even if these regions are presumed to be "causally inert," but, in addition, that there is no reason to treat space-time *as* causally inert because

> a field theory is *most naturally* construed as a theory that ascribes causal properties (electromagnetic field values of various gravitational tensors, etc.) to space-time points. Consequently, any theory in which talk of fields is to be taken seriously (not as a mere shorthand for talk about physical objects) will be most naturally construed as a theory ascribing causal properties to space-time points . . . in such a way that which properties are assigned causally affects the behavior of objects. (p. 70)

Before taking up these considerations, let's rehearse two previously made points: First, it's *never enough* after having established an ontological commitment to something, to simply rest *on that* to *carry along* addi-

---

[61] These claims, by the way, are compatible with both my hand and the table being constituted of such things, and with such constituents existing. But notice that the existence of such constituents is established by thick epistemic access *to them*, not by the mere existence of a true theory that describes the properties of tables and hands in terms of such constituents, because such a theory could prove to be empirically adequate by virtue of its global virtues, without our having determined that *anything* it was quantifier committed to existed.

[62] If space and time themselves don't exist, there's little reason for an ontological commitment to spatial, temporal, or space-time points. Still, for philosophical reasons, some further remarks about spatial, temporal, and space-time points, and the philosophical claims and problems ontological commitments to such things have raised, are in order.

tional ontological commitments that an accepted theory about that thing (merely) posits. To do this is to slip back into a Quinean-style epistemology. Types of posits come with individual requirements on our taking them to really exist. Thus, suppose a way of *construing* (theoretically) a kind K of object (which we *are* committed to) as constituted of, say, objects of a certain sort J is theoretically indispensable. This isn't enough to ontologically commit us to Js.

Second, establishing thick epistemic access to something is more than simply *saying* that causal relations (or worse, some *other* species of relation) *exist* between us and such objects. It must be shown that these relations suffice to enable us to "lock onto" the objects in question in a rich enough way that not only will we be able to determine *properties* of those objects, but that also (in principle) we can find those objects to have properties *at variance with* our theories about what those properties should be. (See chapter 6, esp. conditions (1) – (4) on thick epistemic access.)

Let's turn now to Field's arguments that epistemic access to space-time regions and space-time points isn't as problematical as it is to, say, numbers. (I'll show that Field's considerations are vitiated by the points made in the last two paragraphs.)

Field (1989) stresses the contrast between numbers and space-time regions this way: The idea is that numbers are "outside of space-time" and without causal connections to us. To answer the questions of how we can have any knowledge of numbers or how the word "two" refers to 2 or how a belief about 2 manages to be about 2, "one is going to have to postulate some *aphysical connection*, some *mysterious mental grasping*, between ourselves and the elements of this platonic realm" (p. 68). Such isn't true of space-time regions; here "there are quite unproblematical physical relations, [namely], spatial relations, between ourselves and space-time regions, and this gives us epistemological access to space-time regions. For instance, because of their spatial relations to us, certain space-time regions are close enough to us in space and time to fall within our field of vision."

Field adds later that such regions also

> stand in unproblematic relations (spatiotemporal relations) to physical objects, and this provides us with less direct observational means of knowing about them. No close analogue of this indirect observational access is available for numbers; that is why for instance we can confirm and disconfirm theories about space-time structure by observation, but can't confirm or disconfirm theories about numbers this way. (pp. 68–9)

It should already be clear how this sins against the considerations I've rehearsed, but let's note some details. Contrast space-time regions with electrons, and notice how, in forging thick epistemic access to electrons, a battery of tools (and theoretical considerations) from those of Millikan's until today have been used to empirically verify specific properties of electrons. Not only does this include determining its mass, charge, and so on, but it also includes the quantum-mechanical construal of it as a field: One explores the *geometric distribution* of electron fields around molecules. Contrast this with the *visual* considerations Field invokes on behalf of space-time regions: We don't *see* any of the properties of space-time regions, not their dimensionality, not their curvature, not their composition (out of *whatever*).[63] Indeed, it's bizarre to say we see them at all. *And*, the consideration Field raises about empirically verifying properties of space-time (vs. verifying properties of numbers) is simply a red herring: These properties of space-time are verified only in good Quinean fashion, as indispensable parts of a theory empirically verified as a whole—and *not* because of the presumed empirical relations between us and space-time regions used to forge thick epistemic access to such regions.[64]

But Field may seem to rely on stronger grounds if he is allowed that (1) fields are causal agents (to which, therefore, we *can* forge thick epistemic access) and that (2) such items are "*most naturally* construed" as made up of space-time points or, perhaps more accurately, as simply the exhibition of the various causal properties of space-time points themselves.[65]

Although I've no problem granting an ontological commitment to fields on the grounds of their causal powers, and agree that we do indeed forge thick epistemic access to them, as to, say, electron fields,[66] it simply *doesn't follow* that we can then comfortably take ourselves to be commit-

---

[63] And, as the preceding remarks on Nerlich's example make clear, it's not obvious that we can even make sense of seeing *spatial regions*, vs. seeing *something else* that we (theoretically construe) as spatial regions.

[64] Access to electrons, of course, is deeply embedded in theory, too. But the empirical element of thick epistemic access to the shape of the distributions of electrons around molecules (their probability of location) plays a gigantic role, especially because the formulas that govern this are so intractable.

[65] Field (1989, pp. 70–1) argues against those who would distinguish a field from the space-time it pointwise rests upon. Although the *latter* distinction of space-time from the fields that live on it strikes me as "more natural" than identifying the field with space-time (at least if ordinary practices among physicists are any indication of "naturalness"), I concede this to Field because I argue it won't help anyway.

[66] Do we forge thick epistemic access to *fields*, or is it only that we forge thick epistemic access to electrons that, for theoretical reasons, we *take* to be fields? The former—at least, it's the former to which thick epistemic access is being attempted. Schewe and Stein 2000 write:

> [O]ne phenomenon is well established: Injecting an electron into liquid helium can cause a tiny bubble (approximately 40 angstroms wide) to form around it. Shining light on such bubbles, researchers expected the electrons to escape and zip out of the helium. Not only did they fail to detect such electrons, but other experimenters observed the creation of charged particles, never successfully identified.

The article goes on to summarize a radical explanation by the physicist Humphrey Maris:

> [T]he light causes the electron to break up into two or more pieces, which he terms "electrinos." . . . This split-up, according to Maris, would arise from the quantum abil-

ted to space-time *points*. Granting the indispensability of space-time-point posits in field theory is insufficient to force an ontological commitment to such things.

I hope it's clear that I'm *not* denying an ontological commitment to space-time points simply because they're *extensionless*. That's not a problem—at least not in principle. I can imagine a theory that attributes to genuine point-singularities all sorts of causal powers, and I can imagine—again, in principle—how these powers might be verified empirically by gaining thick epistemic access to such a singularity (or to a collection of them). I can also imagine our being able to verify instrumentally (at least up to whatever measurement thresholds our instruments have achieved) its extensionless status.

But, regardless of whether my imaginings here are physically plausible or not, the point is that nothing like this is going on *in science*; fields are (mathematically) composed of points without these points being instrumentally approached in *any* way (nor is an instrumental approach to such things even being *contemplated*). It's this that tells us that the physics community sees such things as ultrathin.

It might be thought that although I've successfully *denied* Field's reasons for seeing an ontological commitment to space-time points as less problematical than an ontological commitment to numbers, nevertheless I've conceded his point about *space-time regions*. These are epistemically unproblematic because of their identification (which I've agreed with for the sake of argument) with *fields*. Not so. As I'll discuss more fully when I turn to example (f), space-time regions—if seen as *mathematical*—aren't items we have thick epistemic access to. Rather, it's simply that the mathematical nomenclature of space-time regions has been drafted for the characterization of *fields*. But in so doing, it *isn't* that we've gotten thick epistemic access (free of charge, as it were) to mathematical objects because of *their* identification with something nonmathematical; rather, the mathematical posits are *proxying for* real things, and only the latter are what we gain thick epistemic access to.[67]

I want to close this discussion of the ontological status of space-time points by briefly discussing two other ways in which taking such points as real leads to philosophical puzzles or philosophical claims (of a radical sort). Philosophical puzzles can arise with any sort of mathematical abstracta in use in an empirical theory, and mistakenly taken to be real, as I'll indicate in passing.

---

ity of electrons—and all matter—to act in some situations not as particles but as waves. ... Splitting up electron waves also has been demonstrated before . . . , but they ordinarily recombine and get detected as full particles.

[67] For purposes of this example, I've allowed talk of *fields* to be treated as *empirical* language, but, of course, "field" is a mathematical term that is itself drafted to describe something empirical that can't otherwise be characterized in language. See the discussion of (f) later.

I've stressed how the indispensable use of mathematical posits in an empirical theory is (all by itself) insufficient for an ontological commitment to those posits. But there is another side to philosophical construals of the employment of mathematical abstracta in empirical theories: *Alternative* mathematical formulations of what's otherwise the very same empirical theory are usually available. Because these formulations aren't equivalent in the posits they presuppose, this can lead to worries about empirically equivalent theories that are taken as inequivalent in their ontology, and this, in turn, gives rise to purported (and purportedly easy to construct) examples of competing empirical theories the adjudication of which is underdetermined by present (and future) data.

I won't mount an argument against the underdetermination of theory by data doctrine, for to do so calls for much more in the way of detailed analysis than is appropriate now. My only point is that the number of such cases is *greatly* inflated if the presence of a *pure* mathematical posit in an empirical theory (and, of course, its absence in an alternative theory) is used to make such theories distinct in their ontological ("theoretical") commitments. It's also worth adding that the practice of scientists in this respect is at variance with the claims of philosophers plying such underdetermination claims: Scientists routinely take alternative mathematical formulations of the same empirical theories as alternative mathematical formulations of the same empirical theories.[68]

An accessible version of this sort of underdetermination maneuver may be found in Putnam (1976, pp. 130–3), where it's plied against "metaphysical realism" and for "internal realism." The idea is a simple one: It's possible to start with points as primitives and to set-theoretically construct lines and other geometrical objects from those points. It's also possible to start with lines (or some larger geometric object) and construct points (again, set-theoretically) out of these. Mathematically speaking, the two approaches are identical; that is, the *geometrical* (although not set-theoretical) properties of these objects are the same.

Imagine, however, that one sees oneself as committed to geometrical objects (of some sort) because of their indispensable use in one or another scientific theory. Then the following ontological question can be posed: Are there points *really* (and so are lines *really* made out of points)? Or are there lines *really* (and so are points *really* made out of lines)? The "'hard-core' realist," Putnam (1976) writes, "might claim that there is a 'fact of the matter' as to which is true—Story 1 or Story 2" (p. 131).

Ontologically speaking, this concern has bite only if we take the questions of whether there are points or lines (or both) and which are made out of which as *significant* questions about *real things*. This is because

---

[68] E.g., the Schrödinger and Heisenberg representations of quantum mechanics, the Lagrangian and Hamiltonian formulations of Newtonian physics, etc. There are many examples of empirical theories that differ substantially in their mathematical resources but are seen by physicists as, nevertheless, identical theories. The approach to ontology described in part I allows us (in principle) to make sense of the scientific take on such theories.

composition of real things out of other real things isn't a matter of set-theoretical constructions (which can go in any number of ways, depending on sheer mathematical taste) but a matter of how items *are really* put together out of other items and, more important, how the laws that items of one sort obey are compatible with and lead to laws governing how the other sort of items act. That is, if the question of whether points or lines are real is *more* than a mathematical question (which would be answered simply by saying, "It doesn't matter—do whatever is mathematically convenient"), then it's important because how these things are actually put together affects the laws that govern them.

The point can be put graphically: Can we *physically* tear apart lines and find points? Or is it that points are objects composed of lines (so that we find, upon inspection of points—upon forging thick epistemic access to points in a way that allows us to take them apart and be sure what those parts are like—that points are, say, classes of lines, instead)? These considerations are, ultimately, not relevant to the question of how *points* and *lines* are composed of one another because *that* question is just a mathematical one to be stipulated by the mathematical system chosen: Neither points nor lines are anything more than ultrathin. Composition questions, when *real* objects are involved, aren't questions to be dealt with logically or set-theoretically, but ones to be answered by forging thick epistemic access to such objects *and* their parts (as physicists have done with *atoms*) and finding *physically real* methods of tearing things *into* these parts. Although mathematics may be pertinent to the theory of a type of real object (the way that mathematics may be pertinent to any scientific theory), it doesn't follow that questions of the composition of such things are therefore to be answered by the employment of logical, mereological, or set-theoretical tricks of various sorts.

My second example is one I officially take from Zeno's paradoxes of motion.[69] Salmon (and Russell before him) has suggested that the modern invention of calculus has solved Zeno's paradoxes of motion. Here is Salmon (1977):

> The solution of the arrow paradox to which Russell refers[70] has been aptly called "the at-at theory of motion." Using the definition of a mathematical

---

[69] Although I'm not, I should emphatically note, claiming that what I offer *was* a concern of Zeno's.

[70] Russell writes:

> Weirstrass, by strictly banishing all infinitesimals, has at last shown that we live in an unchanging world, and that the arrow, at every moment of its flight, is truly at rest. The only point where Zeno probably erred was in inferring (if he did infer) that because there is no change, therefore the world must be in the same state at one time as at another. This consequence by no means follows. (quoted in Salmon 1970, p. 23)

Salmon describes this remark as somewhat cryptic—but it's quite clear what Russell is implying: that *there is no change*, and that's compatible with things at one moment being one way and at another being another way.

function supplied in the nineteenth century by Cauchy, it is pointed out that the mathematical description of motion is a function that pairs points of space with instants of time. To move from *A* to *B* is simply to occupy the intervening points at the intervening moments. . . . There is no additional question as to how the arrow gets from point *A* to point *B*; the answer has already been given: by being at the intervening points at the intervening moments. Moreover, there is no additional question about how the arrow gets from one intervening point to another; the answer is the same, namely, by being at the points between them at the corresponding moments. And clearly, there can be no question about how the arrow gets from one point to the next, for in a continuum there is no next point. (p. 198)

Although I'm not in a position to say whether this was Zeno's concern or not, still, it seems that if spatial points *are* physically real, then it's appropriate to ask how something moves from one of them to another one of them. The mechanism by which this movement allegedly occurs is bluntly avoided by simply saying, "There is no additional question to be asked"; nevertheless, we'd still like to know how the arrow moves *off* of a point.[71]

That this question has *not* been answered is shown by Russell's remark just quoted (see n. 70), because *he* draws the conclusion that there is no change. This doesn't mean there is "no change" *over time*; what it means is that there is no change *from* one time (point) *to* another. Things just *are* one way at one point (in time) and different *at* another. If this really is the only conclusion that the at-at theory of succession allows about motion, that's a reason to think that Salmon's deflationary construal of the at-at theory blithely leaves out its ominous-looking metaphysical implications.

To deny an ontological commitment to space-time points isn't to open a route to an answer to the question of how motion is possible. What it does instead is eliminate the presence of an irksome (metaphysical) question that we can't otherwise answer.[72]

The reader may be worried: Suppose the right approach *is* that there are no points. Isn't the question of what the mechanism is of the motion of something (an arrow, say) through space still open? I suspect not. Certain Zenonian puzzles about motion that turn on infinite divisibility, such as the tortoise and the hare, *really are* explained by the calculus. But

---

[71] Strictly speaking, it has to be said, Salmon's invocation of the continuum (or, rather, the density) to deflect this question is a *non sequitor*. (Imagine there were only a finite set of points—the problem would remain!)

[72] Notice that, in any case, solutions to Zeno's puzzle like the at-at theory are highly nonempirical. It's not as if we've designed a theory of some sort about spatial and temporal points that we can then try to design an experiment in order to test. It's this *a prioristic* approach to the problem, even on the part of philosophers otherwise quite steeped in scientific method, that indicates that what's really going on is that mathematical abstracta are being pressed into an ontological role they're unequipped to handle.

the arrow problem (how does something move off of a point?) is different: It's due to the presence of a posit that is supposed to play a role in how something moves (it moves *from* point *to* point). Because of its mathematical nature, that movement is impossible. Removing such a posit allows things to simply move through space in a continuous manner (just as you'd hope them to move).[73]

I've now come to the end of my ontological analysis of example (c). Recall what I've tried to show: The application of Ncm to free-wheeling lumps, despite the apparatus of Ncm, commits us to neither forces nor space-time points. Nor, despite the identification (for purposes of application) of cohesive bodies with lumps, does it commit us to either cohesive bodies or material points. We *are* committed to lumps (on grounds apart from the application of Ncm), but we're not committed, by virtue of this application, to lumps (really) being composed of material points.

Examples (d) and (e) don't pose any additional issues, and although (f) does,[74] most of the hard work has already been done in discussing example (c); I'll be brief, therefore, in describing the ontological commitments that arise from the application of Ncm in these cases. For (d) and (e), the situation is the same as it is for (c). That is, that we only use instrumental tools to access the properties of lumps rather than, say, instruments *and* observation, makes no difference to our ontological commitments. A robust enough epistemic link to lumps suffices to determine

---

[73] Of course it's an empirical question whether things *do* move continuously through space or, instead, move in a more discontinuous (quantum-mechanical) way. For that matter, it's an empirical question whether the geometry attributed to space should be the one presupposed by this commonsense analysis of these examples. My only point here is to indicate how the positing of points as *real* creates problems where otherwise no problems would exist.

The sorts of philosophical problems that points create, when we take them to be real, arise, of course, out of the specific mathematical properties of points. Nevertheless, treating abstracta as real can create (different sorts of) philosophical problems, as I showed in the earlier discussion of absolute space.

I should observe that the original Zenonian puzzle is couched not in terms of points (at one of which, say, the tip of the arrow is located) but in terms of the place (a spatial region) that the arrow occupies. Places in those versions of the paradox play the same role that points do in the version of the paradox I presented. We've no more reason to accept an ontological commitment to places than we have to accept one to points. Retaining either ontological commitment forces us to peculiar positions: I've noted Russell's remark about our changeless universe already. Vlastos (1967, p. 375), for another example, supports the category-mistake suggestion that an arrow, during a moment, is neither moving nor not moving: Either claim is a senseless one. It's hard to see why this should be true if a span of time *really is* made up of moments.

[74] Recall what these examples are: In (d) we have epistemic access to the properties of lumps only through instrumental means. In (e) lumps have properties other than those that can be lawfully described by Ncm (such as, say, colors). In (f) there is no antecedent (empirical) language of lumps: The mathematical language of Ncm has been drafted as the only means of description for the items Ncm is applied to in this case.

properties of them in such a way as to take *them* to be ontologically independent, as *real*.

Example (f) only raises a problem if one thinks that an ontological commitment to cohesive bodies, as this term is used when Ncm is applied in this kind of case, brings with it an ontological commitment to the rest of the mathematical apparatus that talk of cohesive bodies essentially occurs within. But this, as I've already shown (see the section on space-time points in chapter 8), isn't the case. "Cohesive body," as a mathematical term, picks out nothing we're ontologically committed to. When Ncm is *applied* in the way it's described in example (f), "cohesive body" does pick out something that we can't otherwise describe. It's fair to say that "cohesive body," used this way, operates in an ambiguous manner: it continues to play its mathematical role within Ncm; but in addition it refers to something ontologically independent, and because it does so, it commits us to that something. The role of "cohesive body" within Ncm is adequately captured by Ncm being true to the empirical phenomena we apply it to; but this much is true of *every* term in Ncm, and isn't relevant to the ontological question whether any of these terms refer. That, in addition, "cohesive body" in that application, refers to something we're ontologically committed to, and yet "space-time point," which occurs within indispensable sentences with the same truth-value status as indispensable sentences in which "cohesive body" occurs, *doesn't* commit us to anything, is simply due to the different roles "cohesive body" and "space-time point" play epistemically: we forge thick epistemic access to "cohesive bodies," but not to space-time points. This is what I mean when I say that the mathematical term "cohesive body" *proxies for something real*.

Why not simply say that "cohesive body" picks out something real? Well, this *can* be said (after all, I've argued it's true); the problem is, as I discussed when I first presented example (f) in chapter 8, that the referential role of "cohesive body" can't be detached from its mathematical role within Ncm. In order to refer, "cohesive body" must largely continue to operate as a mathematical term despite our forging thick epistemic access to cohesive bodies. This is because the instrumental means that attach "cohesive body" (defeasibly) to what it refers to aren't broad enough in how they allow us to thickly access cohesive bodies to allow us to coin a term "cohesive body," and have it stand on its own (referentially) on the basis of that thick access, and apart from the rest of the mathematics of Ncm.

As a result, despite our having a mathematical apparatus that tells us that cohesive bodies are "made of" points, we can wonder if cohesive bodies are indivisible, or if they're composed of (physically real) parts of one sort or another. The mathematical aspects of the "composition" of cohesive bodies, although relevant to the felicity of the application of Ncm to whatever phenomena it's applied to, aren't relevant to the physical question of what cohesive bodies (understood as things *out there* that "cohesive body" applies to) are made of.[75]

One last point about this: Because, in case (f), the terms that we're ontologically committed to arise, as it were, out of a sea of mathematics, the ontological status of terms can shift without the *theory itself* actually changing. We can decide that what's heretofore a piece of pure mathematics actually picks out something "physically real." This phenomenon is inescapable where the ontological commitments of a mathematical application aren't inherited from an antecedently described empirical domain but are directly built up from the mathematical nomenclature itself. Such is the case with our current fundamental physical theories because none (nearly enough) of the new items physics has discovered has antecedent descriptions but instead all arise from instrumental interactions with things described only from within an applied mathematical formalism.

And this brings us to the last promissory note I want to redeem in *this* book. Recall the discussion of (A2) back in chapter 2, when strategies for avoiding the truth of applied mathematics plus empirical science were raised: (A2) was the option of treating *all* the mathematics and the empirical science in a theory in an instrumental way. The problem, we can see now, is that we *don't want* to treat all the mathematics and empirical science (in case (f)) as false: Instrumental access to mathematically described items suffices (under certain circumstances) to take those items as ontologically independent of us, and (some of) the implications of the theory about those entities are things we want to take as *true*, and as true of *them*.

---

[75] It seems to me that in practice physicists can be clear on this distinction, even when they use *obviously* misleading language. As an illustration, consider the following remarks of Gell-Man (1997):

> I employed the term "mathematical" for quarks that would not emerge singly and "real" for quarks that would.... [T]o illustrate what I meant by "mathematical", I gave the example of the limit of infinite mass and infinite binding energy.
> Later, in my introductory lecture at the 1966 International Conference on High Energy Physics in Berkeley, I improved the characterization of mathematical quarks by describing them in terms of the limit of an infinite potential, essentially the way that confinement is regarded today. Thus what I meant by "mathematical" for quarks is what is now generally thought to be both true and predicted by QCD [quantum chromodynamics]. Yet, up to the present, numerous authors keep stating or implying that when I wrote that quarks were likely to be "mathematical" and unlikely to be "real", I meant that they somehow weren't there. Of course I meant nothing of the kind. (p. 626)

Conclusion

G. E. Moore (1939) once wrote (about a much shorter work than this!) that "I have only, at most, succeeded in saying a very small part of what ought to be said about it" (p. 126). I'm in a similar position despite having written so much more, now and before, on the topics broached in this book. Despite this, I like the idea of leaving future work both for myself, and for others. I want, by way of conclusion, to summarize what I hope I've accomplished in this book, and to sketchily indicate what's left for the future.

Despite Quine's widespread influence, there are very few of his claims that don't continue to be highly controversial. Indeed, on many issues, almost everyone is convinced he is wrong—the disagreement comes in on the question of how.[1] His influence, that is, has largely been in setting the tone and framework for the debate on a number of central issues rather than in providing a philosophical foundation that other philosophers have then used as a springboard for their own work.

This is *not* true of Quine's criterion for what a discourse is committed to, however. Even those philosophers he most decisively broke with on other issues (e.g., Carnap) found themselves embracing the letter of his view of how ontological commitment is to be identified, although Carnap, anyway, deflated the metaphysical implications Quine drew from that view. In part the nearly universal explicit (and implicit) acceptance of

---

[1] This is my take on how the profession currently feels about indeterminacy of translation, for example.

Quine's criterion was because, as Quine himself realized (and was modestly willing to point out more than once), something like his criterion had already been at work among philosophers.[2]

Thus, I recognized (back in 1992) how radical the views were that I was moving toward when I drew the conclusion that the philosophical arguments Quine had used to establish his criterion were not compelling, and that there is room in logical space for a position that simultaneously takes a discourse to be true (which I eventually recognized was something we *had* to do in the case of applied mathematics and empirical scientific doctrine) and yet, despite existential quantification, still allows us to deny ontological commitment to the posits in that discourse. In a series of articles (1997a, 1997b, 1998), I argued for the philosophical indeterminacy of a criterion for what a discourse commits us to, in the sense that no good philosophical arguments (or technical ones, for that matter) are available to establish such a criterion, and a similarly (linked) philosophical indeterminacy for a criterion for what exists.

I've not deserted the general position of those essays, but I *have* found a way of arguing both for a criterion for what a discourse commits us to, and for a criterion for what exists, by recognizing how ordinary citizens of our epistemic and linguistic community take themselves to be committed to something, and what the methods that lead to commitment to something indicate *about* that something. The resulting criteria are not criteria that we as a community *must* adopt on pain of incoherence (one can easily imagine communities that don't share our criteria): Our criteria are simply the criteria we *have* adopted.

These linked criteria (something that exists is "ontologically independent"; a discourse is committed to all items that fall under the "ontological independence" *predicate*) replaces the relatively straightforward Quinean syntactic criterion with a semantic one. As a result, our metaphysical lives become much harder than they appeared to be, given Quine's approach. I describe Quine's criterion as *syntactic* in the sense that, *once a discourse has been regimented*, one need only look for implied sentences with a certain syntactic structure (they begin with an existential quantifier). But once this criterion is replaced with a predicate-style one, determining what falls *under* that predicate is no longer a matter of the *grammar* of the regimentation at all.

Instead, we must look beyond the letter of the theory in question to its application to the world. We must see exactly how, in a sense, that theory is attached to the world. Those posits of the theory that *are meant to be* ones that are secured to the world via a special type of causal relation from us to the world—a causal relation I've described as thick epistemic access—are items that we take to exist. Additional items that are theoreti-

---

[2] I think, for example (as I've tried to illustrate), that, in some form, it was a powerful factor in the substantivalist/relationist debate over space and time—and this is so right from the beginning of that debate.

cally tied to the items just mentioned in a tight enough way are also items we take to exist.

The remaining posits of the theory subside into ultrathin status. We don't take them to exist in any sense at all; they're part of the mathematical formalism that facilitates the application of our theories to the world.

It's in this sense, and only in this sense, that I've made a case for *nominalism*. I've not done so in the programmatic sense of presenting a project of rewriting our scientific theories so that the quantifiers in their regimentations no longer range over abstracta, but (1) in the sense of providing an argument that mathematical abstracta (in pure mathematics) aren't ontologically independent and thus aren't items that we (should) take ourselves to be committed to, and (2) in the sense of providing an approach to the evaluation of scientific theories, in order to sort posits so that we can recognize which ones are ultrathin and thus not items we're committed to.

The exercise of examining a particular scientific theory (and its application) in order to determine which posits *are* the ones we mean to exist isn't an easy or straightforward activity. In part this is because even if the language describing a particular posit of a scientific theory is *entirely mathematical*, it doesn't follow that the posit itself is ultrathin. Mathematical posits (from the point of view of pure mathematics) can *proxy for* items that are ontologically independent of us and that therefore, despite their mathematical clothing (within our theories), are items we're committed to.

We must also be careful in another way. In the application of a theory, we may treat ultrathin posits as items that are causally efficacious; this is done, for example, with centers of mass and, more generally, with point-masses. But these items are *not* genuinely seen as things with causal powers because they're not taken to exist in any sense at all. And the proof that this is our attitude is the lack of any desire (on the part of scientists) to forge thick epistemic access to such objects in order to verify they have the causal powers attributed to them (just as with space-time points).

I exhibited the exercise of evaluating the posits of a scientific theory at fair length in the case of a small piece of Newtonian physics: I showed how various sorts of applications of the mathematical system which I called *Newtonian cohesive-body mathematics* (Ncm) should be evaluated in terms of what their ontology commits us to.

That certain posits prove to be items we're not committed to, given the successful application of a particular theory doesn't mean they'll continue to have that ontological status later. For example, that, in the Newtonian context, the presence of fields as posits in the theory—indeed, the couching of that theory entirely in terms of fields—doesn't mean we're committed to such, is not a lesson that need be true of later theories. This is not merely because of the fact that action-at-distance, in other words, the instantaneous impact of the field throughout space, is replaced by a field that travels at a finite speed. For the latter is simply a change in the properties of a posit (the way that the replacement of a flat space-time by a curved space-time is). It's the (intended) *application* of the theory that tells us whether the posit is to be taken as existing or not. So too, for example, the fact that *energy* is located in space-time itself in general relativity is insufficient for a commitment to that space-time. If, however, thick epistemic access is forged to parts of space-time, then not merely has that posit changed its properties, but its *ontological* status within the theory has changed, as well, because we're attempting thick epistemic access to *it*.

There is a sense, however, in which what I'm offering *is* an ontological program, although in a much weaker sense than are nominalist programs that undertake to rewrite all of science. This is that the ontological views I've argued for in this book are part of the *ordinary* take on ontology, and there is no guarantee that the results they dictate sit easily with contemporary science. That is, the results of contemporary science may force us to revise how we think about *ontology*, even in the most fundamental ways we currently do so. (We can be wrong about *anything*, and in no way are philosophical doctrines immune to surprising developments in science.)

I noted earlier that it's not easy to determine what the posits of a theory are that we take to exist, and I gave one reason for that, namely, that when posits are drawn from a mathematical formalism, it's not given in that formalism which ones (if any) aren't ultrathin. Here is another: Our view of thick epistemic access is itself part of our scientific worldview. Thus, it's possible to be theoretically obscure what sort of things we have (or are trying to have) thick epistemic access *to*. I think, for this reason, the ontological evaluation of, for example, the quantum-mechanical formalism, is highly unobvious.[3]

I also had to set aside any detailed discussion of the special sciences except for the occasional, and probably obscure, hint (see chapter 7, n. 20). There are fascinating issues about how the posits in such sciences work, and how we should sort out which of those we take to exist and which not.

Indeed, when I first set out to write this book, I was intending to discuss issues about the special sciences, especially linguistics, because of

---

[3] Different views of the quantum-mechanical formalism turn in part on different views of which aspects of that formalism should be taken as *real*. Some of the wilder interpretations turn on taking the complex wave function itself as existing (on the grounds, usually, that the laws about that wave function *are* true). Although the ontological views I've argued for in this book don't allow us to take the just mentioned ontological view of the wave function, there is still a great deal of uncertainty about exactly what aspects of the quantum-mechanical formalism should be taken as items we're to gain thick epistemic access to (and therefore items we take to exist). Recent experiments apparently illustrating thick epistemic access to superpositions of a particle may indicate that a purely mathematical interpretation of the *probabilistic field* is not tenable.

the connections I saw between this network of problems and, in particular, the rule-following problem. I promised a discussion of my take on the rule-following problem back in 1994, and during the course of writing this book regretfully found myself having to put it off yet a third time. One cannot do everything at once.

References

Achinstein, Peter, 1965. The problem of theoretical terms. In *Readings in the philosophy of science* (1970), ed. Baruch A. Brody, 234–50. New York: Prentice-Hall.

Adair, R. K., 1991. Quarks. In *Encyclopedia of physics*, 2nd ed., ed. Rita G. Lerner and George L. Trigg, 1001–1004. New York: VCH Publishers.

Armstrong, D. M., 1978. *Universals and scientific realism*. Cambridge: Cambridge University Press.

Austin, J. L., 1962. *Sense and sensibilia*. Oxford: Oxford University Press.

Azzouni, Jody, 1994. *Metaphysical myths, mathematical practice: the ontology and epistemology of the exact sciences*. Cambridge: Cambridge University Press.

———, 1997a. Applied mathematics, existential commitment and the Quine-Putnam indispensability thesis. *Philosophia Mathematica* 5,3:193–209.

———, 1997b. Thick epistemic access: distinguishing the mathematical from the empirical. *Journal of Philosophy* 94:472–84.

———, 1998. On "on what there is." *Pacific Philosophical Quarterly* 79,1:1–18.

———, 1999. Review of Michael D. Resnik's "mathematics as a science of patterns." *Journal of Symbolic Logic* 64:922–3.

———, 2000a. Applying mathematics: an attempt to design a philosophical problem. *Monist* 83,2:209–27.

———, 2000b. *Knowledge and reference in empirical science*. London: Routledge.

———, 2000c. Stipulation, logic, and ontological independence. *Philosophia Mathematica* 8,3:225–43.

———, 2001. Truth via anaphorically unrestricted quantifiers. *Journal of Philosophical Logic* 30:329–54.

———, 2003. Individuation, causal relations, and Quine. In *Meaning*, ed. Mark Richard, 197–219. Oxford: Blackwell.

———, forthcoming(a). Proof and ontology in mathematics. In *Proceedings of new trends in the history and philosophy of mathematics at Roskilde University*, summer 1998.

———, forthcoming(b). Tarski, Quine, and the transcendence of the vernacular "true." *Synthese*.

Barbour, Julian, 1999. *The end of time: the next revolution in physics*. Oxford: Oxford University Press.

Barcan-Marcus, Ruth, 1971. Quantification and ontology. In *Modalities* (1993), 76–87. Oxford: Oxford University Press.

Belot, G., and Earman, J., 1999. From physics to metaphysics. In *From physics to philosophy*, ed. J. Butterfield and C. Pagonis, 166–86. Cambridge: Cambridge University Press.

Bennett, Jonathan, 1969. Real. In *Symposium on J. L. Austin*, ed. K. T. Fann, 267–83. London: Routledge and Kegan Paul.

Bills, Robert E., 1981. Dimensions of the dilemma. In *Self concept: advances in theory and research*, ed. K. Gergen, M. Lynch, and A. Norem-Hebeisen, 17–25. Cambridge, Mass.: Ballinger.

Boolos, George, 1998. Must we believe in set theory? In *Logic, logic, and logic*, ed. Richard Jeffrey, 120–32. Cambridge, Mass.: Harvard University Press.

Brown, Laurie M., Michael Riordan, Max Dresden, and Lillian Hoddeson, 1997. The rise of the standard model: 1964–1979. In *The rise of the standard model: particle physics in the 1960s and 1970s*, ed. Lillian Hoddeson, Laurie Brown, Michael Riordan, and Max Dresden, 3–35. Cambridge: Cambridge University Press.

Burgess, John, and Gideon Rosen, 1997. *A subject with no object*. Oxford: Oxford University Press.

Carnap, Rudolf, 1939. *Foundations of logic and mathematics*. International Encyclopaedia of Unified Sciences, vol. 1, no. 3. Chicago: University of Chicago Press.

———, 1950. Empiricism, semantics, and ontology. In *Meaning and necessity* (1956), 205–21. Chicago: University of Chicago Press.

Cartwright, Nancy, 1983. *How the laws of physics lie*. Oxford: Oxford University Press.

———, 1989. *Nature's capacities and their measurement*. Oxford: Oxford University Press.

Cartwright, Richard, 1954. Ontology and the theory of meaning. In *Philosophical essays* (1987), 1–12. Cambridge, Mass.: MIT Press.

Chang, C. C., and H. Jerome Keisler, 1973. *Model theory*. Amsterdam: North Holland.

Cohen, I. Bernard, 1999. A guide to Newton's *Principia*. In *The* Principia: *mathematical principles of natural philosophy*, trans. I. Bernard Cohen and Anne Whitman, 1–370. Berkeley: University of California Press.

Collins, Arthur, 1998. On the question "do numbers exist?" *Philosophical Quarterly* 48,190:23–36.

Colyvan, Mark, 2001. *The indispensability of mathematics*. Oxford: Oxford University Press.

Cosmides, Leda, John Tooby, and Jerome H. Barkow, 1992. Introduction: evolutionary psychology and conceptual integration. In *The adapted mind: evolutionary psychology and the generation of culture*, ed. Jerome H. Barkow, Leda Cosmides, and John Tooby, 3–15. New York: Oxford University Press.

Cunningham, Walter R., and Adrian Tomer, 1990. Intellectual abilities and age: concepts, theories and analyses. In *Aging and cognition, mental processes, self-awareness and interventions*, ed. Eugene A. Lovelace, 379–406. Amsterdam: North-Holland.

Dainton, Barry, 2001. *Time and space.* Chesham, UK: Acumen Publishing.

Davidson, Donald, 1965. Theories of meaning and learnable language. In *Inquiries into truth and interpretation* (1986), 3–15. Oxford: Oxford University Press.

———, 1967. Causal relations. In *Essays on actions and events* (1986), 149–62. Oxford: Oxford University Press.

———, 1977. The method of truth in metaphysics. In *Inquiries into truth and interpretation* (1986), 199–214. Oxford: Oxford University Press.

———, 1999. Reply to J.J.C. Smart. In *The philosophy of Donald Davidson*, ed. Lewis Edwin Hahn, 123–5. Chicago: Open Court.

Deutsch, Harry, 2000. Making up stories. In *Empty names, fiction and the puzzles of non-existence*, ed. Anthony Everett and Thomas Hofweber, 149–81. Stanford, Calif.: CSLI Publications.

Donnellan, Keith, 1974. Speaking of nothing. *Philosophical Review* 83:3–32.

Dummett, M. A. E., 1963. Realism. In *Truth and other enigmas*, 145–65. Cambridge, Mass.: Harvard University Press.

Einstein, Albert, 1916. The foundations of the general theory of relativity. In *The principle of relativity* (1923), trans. W. Perrett and G. B. Jeffery, 109–64. New York: Dover Publications.

Enderton, Herbert B., 1972. *A mathematical introduction to logic.* New York: Academic Press.

Feynman, Richard, Robert B. Leighton, and Matthew Sands, 1963. *The Feynman lectures on physics.* Reading, Mass.: Addison-Wesley Publishing.

Field, Hartry, 1972. Tarski's theory of truth. *Journal of Philosophy* 69,13:347–75.

———, 1980. *Science without numbers.* Princeton, N.J.: Princeton University Press.

———, 1989. *Realism, mathematics and modality.* Oxford: Blackwell.

Folger, Tim, 2000. From here to eternity. *Discover* 21,12:54–61.

Franklin, Allen, 1993. *The rise and fall of the fifth force.* New York: American Institute of Physics.

Friedman, Michael, 1983. *Foundations of space-time theories: relativistic physics and philosophy of science.* Princeton, N.J.: Princeton University Press.

Galison, Peter, and Bruce Hevley, 1992. *Big science: the growth of large-scale research.* Stanford, Calif.: Stanford University Press.

Gell-Mann, Murray, 1997. Quarks, color, and QCD. In *The rise of the standard model: particle physics in the 1960s and 1970s*, ed. Lillian Hoddeson, Laurie Brown, Michael Riordan, and Max Dresden, 625–33. Cambridge: Cambridge University Press.

Gottlieb, Dale, 1980. *Ontological economy: substitutional quantification and mathematics.* Oxford: Oxford University Press.

Hacking, Ian, 1981. Do we see through a microscope? In *Images of science: essays on realism and empiricism, with a reply from Bas C. van Fraassen* (1985), ed. Paul M. Churchland and C. A. Hooker, 132–52. Chicago: University of Chicago Press.

———, 1983. *Representing and intervening.* London: Cambridge University Press.

Hale, Bob, 1987. *Abstract objects*. Oxford: Blackwell.
Hatfield, Gary, 1990. Metaphysics and the new science. In *Reappraisals of the scientific revolution*, ed. David C. Lindberg and Robert S. Westman, 93–166. Cambridge: Cambridge University Press.
Hempel, Carl, 1948. Studies in the logic of explanation. In *Aspects of scientific explanation and other essays in the philosophy of science* (1965), 3–51. New York: Free Press.
Hofweber, Thomas, 2000. Quantification and non-existent objects. In *Empty names, fiction and the puzzles of non-existence*, ed. Anthony Everett and Thomas Hofweber, 249–73. Stanford, Calif.: CSLI Publications.
Hume, David, 1739. *A treatise of human nature*. New York: Dolphin Books (1961).
Jackson, Frank, 1976. The existence of mental objects. *American Philosophical Quarterly* 12:33–40. Reprinted in *The nature of mind* (1991), ed. David Rosenthal, 385–91. Oxford: Oxford University Press.
Kant, Immanuel, 1783. *Prolegomena of any future metaphysics that will be able to come forth as science* (1997), ed. Gary Hatfield. Cambridge: Cambridge University Press.
———, 1787. *Critique of pure reason* (1929), trans. Norman Kemp Smith. New York: St. Martin's Press.
Katz, Jerrold J., 2002. Mathematics and metaphilosophy. *Journal of Philosophy* 94:362–90.
Kim, Jaegwon, 1993. *Supervenience and mind: selected philosophical essays*. Cambridge: Cambridge University Press.
Kitcher, Philip, 1981. Explanatory unification. *Philosophy of Science* 48:507–31.
Koslow, Arnold, 2000. Ontology and scientific laws. In *III. International Ontology Congress (San Sabastian, 1998), Physis: from Greek thought to quantum mechanics (UNESCO)*, ed. V. Gomez Pin, 61–72. Muelle de la Merced, 3–2^0, Izda 48003, Bilbao: Imprenta Luna.
Krantz, David H., R. Duncan Luce, Patrick Suppes, and Amos Tversky, 1971. *Foundations of measurement*, vol. 1. New York: Academic Press.
Kuhn, Thomas S., 1970. *The structure of scientific revolutions*. Chicago: University of Chicago Press.
Landau, L. D., and E. M. Lifshitz, 1977. *Quantum mechanics*, 3rd ed. New York: Pergamon Press.
Landesman, Charles, 1975. Thought, reference and existence. *Southern Journal of Philosophy*, 13,4:449–58.
Lethem, Jonathan, 1995. *Amnesia moon*. New York: Harcourt Brace.
Levin, Michael, 1984. What kind of explanation is truth? In *Scientific realism*, ed. Jarrett Leplin, 124–39. Berkeley: University of California Press.
Levy, Azriel, 1979. *Basic set theory*. Berlin: Springer-Verlag.
Lewis, David, 1986. *On the plurality of worlds*. Oxford: Blackwell.
Maddy, Penelope, 1992. Indispensability and practice. *Journal of Philosophy 89*: 275–89.
———, 1994. Taking naturalism seriously. In *Logic, methodology and philosophy of science IX*, ed. D. Prawitz, B. Skyrms, and D. Westerstahl, 383–407. Amsterdam: Elsevier.
———, 1997. *Naturalism in mathematics*. Oxford: Oxford University Press.
Maudlin, T., 1993. Buckets of water and waves of space: why spacetime is probably a substance. *Philosophy of Science 60*:183–203.

Maxwell, Grover, 1962. The ontological status of theoretical entities. In *Readings in the philosophy of science*, ed. Baruch A. Brody, 224–33. New York: Prentice-Hall.

Moore, G. E., 1939. Proof of an external world. In *Philosophical papers* (1962), 126–48. New York: Collier Books.

Nerlich, Graham, 1994a. *The shape of space*, 2nd ed. Cambridge: Cambridge University Press.

———, 1994b. *What spacetime explains: metaphysical essays on space and time.* Cambridge: Cambridge University Press.

Newton, Isaac, 1726. *The* Principia*: mathematical principles of natural philosophy* (1999), trans. I. Bernard Cohen and Anne Whitman. Berkeley: University of California Press.

Noll, Walter, 1974. *The foundations of mechanics and thermodynamics: selected papers.* Berlin: Springer-Verlag.

Oddie, G., 1982. Armstrong on the Eleatic Principle and abstract objects. *Philosophical Studies* 41:285–95.

Parsons, Terence, 1980. *Nonexistent objects.* New Haven, Conn.: Yale University Press.

Perrin, Jean Baptiste, 1990. *Atoms.* Woodbridge, Conn.: Oxbow.

Purcell, Edward, 1985. *Electricity and magnetism. Berkeley physics course*, vol. 2. New York: McGraw-Hill.

Putnam, Hilary, 1962. What theories are not. In *Mathematics, matter and method: philosophical papers*, vol. 1 (1979), 215–27. Cambridge: Cambridge University Press.

———, 1975. What is mathematical truth? In *Mathematics, matter and method: philosophical papers*, vol. 1 (1979), 60–78. Cambridge: Cambridge University Press.

———, 1976. Realism and reason. In *Meaning and the moral sciences*, 123–40. London: Routledge and Kegan Paul.

———, 1987. *The many faces of realism.* LaSalle, Ill.: Open Court.

Psillos, Stathis, 1999. *Scientific realism: how science tracks truth.* London: Routledge.

Quine, W. V., 1941. Whitehead and modern logic. In *Selected logic papers*, enl. ed., 3–36. Cambridge, Mass.: Harvard University Press.

———, 1948. On what there is. In *From a logical point of view* (1980), 1–19. Cambridge, Mass.: Harvard University Press.

———, 1950. Identity, ostension, and hypostasis. In *From a logical point of view* (1980), 65–79. Cambridge, Mass.: Harvard University Press.

———, 1951. Two dogmas of empiricism. In *From a logical point of view* (1980), 20–46. Cambridge, Mass.: Harvard University Press.

———, 1953a. Mr. Strawson on logical theory. In *The ways of paradox and other essays* (1976, rev. enl. ed.), 137–57. Cambridge, Mass.: Harvard University Press.

———, 1953b. Reference and modality. In *From a logical point of view* (1980), 139–59. Cambridge, Mass.: Harvard University Press.

———, 1955. Posits and reality. In *The ways of paradox and other essays* (1976, rev. enl. ed.), 246–54. Cambridge, Mass.: Harvard University Press.

———, 1960. *Word and object.* Cambridge, Mass.: MIT Press.

———, 1969. Existence and quantification. In *Ontological relativity and other essays* (1969), 91–113. New York: Columbia University Press.

———, 1972a. *Methods of logic.* New York: Holt, Rinehart and Winston.

———, 1972b. The variable. In *The ways of paradox and other essays* (1976, rev. enl. ed.), 272–82. Cambridge, Mass.: Harvard University Press.

———, 1975. On the individuation of attributes. In *Theories and things* (1981), 100–12. Cambridge, Mass.: Harvard University Press.

———, 1978. Goodman's ways of worldmaking. In *Theories and things* (1981), 96–9. Cambridge, Mass.: Harvard University Press.

———, 1980. *Elementary logic*, rev. ed. Cambridge, Mass.: Harvard University Press.

———, 1981a. Things and their place in theories. In *Theories and things* (1981), 1–23. Cambridge, Mass.: Harvard University Press.

———, 1981b. What price bivalence? In *Theories and things* (1981), 31–7. Cambridge, Mass.: Harvard University Press.

———, 1981c. Responses. In *Theories and things* (1981), 173–86. Cambridge, Mass.: Harvard University Press.

———, 1986. *Philosophy of logic*, 2nd ed. Cambridge, Mass.: Harvard University Press.

Resnik, Michael D., 1995. Scientific and mathematical realism: the indispensability argument, *Philosophia Mathematica* 3:166–74.

———, 1997. *Mathematics as a science of patterns.* Oxford: Oxford University Press.

Routley, Richard, 1980. *Exploring Meinong's jungle and beyond: an investigation of noneism and the theory of items*, interim ed. Philosophy department monograph 3. Canberra: Research School of Social Sciences, Australian National University.

Russell, Bertrand, 1912. *Problems of philosophy.* Indianapolis, Ind.: Hackett Publishing.

———, 1919. *Introduction to mathematical philosophy.* London: Allen and Unwin.

Salmon, Nathen, 1998. Nonexistence. *Nous* 32,3:277–319.

Salmon, Wesley, 1970, ed. *Zeno's paradoxes.* Indianapolis, Ind.: Bobbs-Merrill.

———, 1977. An "at-at" theory of causal influence. In *Causality and explanation*, 193–9. New York: Oxford University Press.

Scheffler, Israel, and Noam Chomsky, 1958–59. What is said to be. *Proceedings of the Aristotelian Society* 59:71–82.

Schewe, Phillip F., and Ben Stein, 2000. Can electrons split up into smaller fragments? *American Institute of Physics Bulletin of Physics News*, no. 501, Sept. 7.

Schiffer, Stephen, 1996. Language-created language-independent entities. *Philosophical Topics* 24,1:149–67.

Shapiro, Stewart, 1991. *Foundations without foundationalism.* Oxford: Oxford University Press.

Sklar, Lawrence, 1976. *Space, time and spacetime.* Berkeley: University of California Press.

Smart, J. J. C., 1999. Correspondence, coherence, and realism. In *The philosophy of Donald Davidson*, ed. Lewis Edwin Hahn, 109–22. Chicago: Open Court.

Smith, George, 2001. The Newtonian style in book 2 of the *Principia*. In *Isaac Newton's natural philosophy*, ed. J. Z. Buchwald and I. B. Cohen, 249–313. Cambridge, Mass.: MIT Press.

———, 2002. The methodology of the *Principia*. In *The Cambridge companion* to Newton, ed. I. B. Cohen and G. E. Smith, 138–173. Cambridge: Cambridge University Press.

Sober, Elliott, 1993. Mathematics and indispensability. *Philosophical Review* 102: 35–57.

Stein, Howard, 1967. Newtonian space-time. *Texas Quarterly* 10:174–200.

Steiner, Mark, 1989. The application of mathematics to natural science. *Journal of Philosophy* 83:249–64.

———, 1998. *The applicability of mathematics as a philosophical problem*. Cambridge, Mass.: Harvard University Press.

Strasberg, M., 1991. Acoustical measurements. In *Encyclopedia of physics*, ed. Rita G. Lerner and George L. Trigg, 9–13. New York: VCH Publishers.

Sussman, Gerald Jay, and Wisdom, Jack, 1992. Chaotic evolution of the solar system. *Science* 257:56–62.

Truesdell, C., 1979. Statistical mechanics and continuum mechanics. Reprinted in *An idiot's fugitive essays on science* (1984), 72–9. New York: Springer-Verlag.

———, 1980–1. Suppesian stews. Reprinted in *An idiot's fugitive essays on science* (1984), 503–79. New York: Springer-Verlag.

———, 1991. *A first course in rational continuum mechanics*, 2nd ed. Boston: Academic Press.

Truesdell, C., and K. R. Rajagopal, 2000. *An introduction to the mechanics of fluids*. Boston: Birkhäuser.

Urmson, J. O., 1976. Fiction. *American Philosophical Quarterly* 13:153–7.

van Fraassen, Bas C., 1980. *The scientific image*. Oxford: Oxford University Press.

van Inwagen, Peter, 2000. Quantification and fictional discourse. In *Empty names, fiction and the puzzles of non-existence*, ed. Anthony Everett and Thomas Hofweber, 235–47. Stanford, Calif.: CSLI Publications.

Vinueza, Adam, 2001. Realism and mind independence. *Pacific Philosophical Review* 82:51–70.

Vlastos, Gregory, 1967. Zeno of Elea. In *The encyclopedia of philosophy*, vol. 8, ed. Paul Edwards, 369–79. New York: Macmillan.

Walton, Kendall L., 1990. *Mimesis as make-believe*. Cambridge, Mass.: Harvard University Press.

———, 2000. Existence as metaphor? In *Empty names, fiction and the puzzles of non-existence*, ed. Anthony Everett and Thomas Hofweber, 69–94. Stanford, Calif.: CSLI Publications.

Weinberg, Steven, 1993. *Dreams of a final theory*. New York: Vintage Books.

Weiskrantz, L., 1986. *Blindsight: a case study and implications*. Oxford: Oxford University Press.

Westfall, Richard S., 1980. *Never at rest: a biography of Isaac Newton*. Cambridge: Cambridge University Press.

Wittgenstein, Ludwig, 1922. *Tractatus logico-philosophicus* (1961), trans. D. F. Pears and B. F. McGuinness. London: Routledge.

Wright, Crispin, 1992. *Truth and objectivity*. Cambridge, Mass.: Harvard University Press.

Yablo, Stephen, 1998. Does ontology rest on a mistake? *Proceedings of the Aristotelian Society* 72(suppl). 229–62.

———, 2000.Apriority and existence. In *New essays on the a priori*, ed. Paul Boghossian and Christopher Peacocke, 197–228. Oxford: Oxford University Press.

Index

"absolute," 199 n. 40
Achinstein, Peter, 39 n. 20
Adair, R. K., 133
adverbial and adjectival inferences, ontological implications of, 58, 69–70
Alexander, H. G., 199–200
algorithmic systems, 160 n. 2, 161–162
applying mathematics
    explanation for success in, 165, 174–180, 195
    inexactness of, 162–164
approximating theories, 44
Armstrong, David, 149 n. 8
Austin, J. L., 119 n. 10, 119 n. 11

Babour, Julian, 8
Ballaguer, Mark, 95 n. 26
Barcan-Marcus, Ruth, 54 n. 9, 65 n. 31
belief
    idealizations of, 9 n. 13, 41n
    manifestion condition on, 36–37, 39–40, 41, 44–45
Belot, G., 8 n. 9

Bennett, Jonathan, 119 n. 10
Bills, Robert E., 132 n
black box objection, the, 144–147, 210
Boolos, George, 119–120
Boyd, Richard. *See* Boyd-Putnam "no miracle argument"
Boyd-Putnam "no miracle argument," 145–146
Brown, Laurie, M., 133, 148 n. 5, 151 n. 13
Burgess, John, 30 n. 2

calculational shortcuts, 23
Carnap, Rudolf, 3, 4n, 6–7, 51, 222
Cartwright, Nancy, 29n, 30–31, 39 n. 19, 40–41, 44, 45 n. 24, 83 n. 4, 84 n. 7, 155n, 177
Cartwright, Richard, 50 n. 2
Chang, C. C., 18 n. 8
Chomsky, Noam, 50 n. 2
Clarke, Samuel, 196
co-extensionality*, 70–71, 89–90, 120n, 155. *See also* co-reference*; opacity

Cohen, I. Bernard, 186 n. 11, 186–187 n. 12, 191n, 204
coherentist epistemology, 109, 144–145
cohesive bodies, ontological status of, 192–193, 220–221
Collins, Arthur, 126n
Colyvan, Mark, 62 n. 26, 76 n. 51, 149 n. 8, 155–157
confirmation holism, 27–28, 76–77, 118
constant angular velocity, 207–208, 208n
constitutive relations, 169 n. 14, 170, 182 n. 1, 182 n. 2, 184, 185, 187 n. 14, 188 n. 16, 192, 202–203, 205
  characterization of, 168
co-reference*, 70n, 78, 79n, 92, 122. See also extension*
  definition of, 61–62
Cosmides, Leda, 154 n. 20
criterion for what a discourse is committed to. See also Quine, his criterion for what a discourse is committed to
  indeterminateness of, 98–99
criteria for what exist, 5, 50, 149
  indeterminateness of, 5–6, 98–99
criterion transcendence, 126, 159, 201n. See also vernacular "exists," as a criterion transcendent term
Cunningham, Walter R., 132n

Dainton, Barry, 207 n. 56, 208n, 209, 210
Davidson, Donald, 17 n. 5, 54 n. 9, 57–58, 144
decision procedure, 17–18
deduction, idealizations of, 17, 22
deflating ontology, 6–7
Deutsch, Harry, 71, 95 n. 26
Donnellan, Keith, 74 n. 44
Duhem, Pierre, 34 n. 11, 45 n. 24
Dylan, Bob, 38

Earman, John, 8 n. 9
Einstein, Albert, 47, 196, 204 n. 50, 205–207

eleatic principle, the, 149 n. 8, 150, 154–155, 154 n. 21, 156 n. 23
Enderton, Herbert, 17 n. 5
epistemic burdens, characterization of, 127
Euclidean geometry, 160 n. 2, 161, 162, 168–169
  as a backdrop for Ncm, 166–167
epistemic role obligation, 136, 143
  characterization of, 135
existence. See also existence predicate; vernacular "exists."
  necessary and sufficient conditions on, 87–98, 112 (See also ontological independence)
existence predicate, 4, 49, 52 n. 6, 79, 80, 82, 84, 88–89, 114, 127, 223
extension*, 149

facticity of explanation, 45 n. 24
false theories, value of, 35–36
Feynman, Richard, 23 n. 18, 47
fictional entities, nonexistence of, 71–72, 83, 93–98
Field, Hartry, 30, 43, 62 n. 26, 83 n. 5, 84 n. 7, 99 n. 33, 194 n. 32, 212–215, 214 n. 65
finite axiomatizability, 18, 19
  definition of, 17 n. 4
Folger, Tim, 8 n. 9
folk ontology, 4, 7–9, 51, 53, 77–78, 116 n. 3, 117–122. See also vernacular "there is," evaluation of ontological commitments of; vernacular "exists"
  fallibility of, 225
forces
  additivity of, 184 n. 5
  as an inevitable manner of speaking, 188
  genuine vs. pseudo, 186–187, 189
  explanatory role of, 185–186
  fundamental vs. nonfundamental, 184–185
  Newton's commitment to, 186 n. 12
  ontological dependence on, 190–191
  thick epistemic access to, 185n

Franklin, Alan, 132n
Friedman, Michael, 146–147, 166 n. 5, 197 n. 35, 197 n. 38, 198 n. 39, 207

Galison, Peter 129 n. 7
Gell-Man, Murray, 221n
global pretence, threat of, 75–76
Gottlieb, Dale, 50 n. 2, 76 n. 50
Grice, Paul, 66
gross regularities, 131
grounding, 133, 134, 135, 138, 141, 143
    characterization of, 129

Hacking, Ian, 29n, 44, 83 n. 4, 84 n. 7, 132n
Hale, Bob, 77
Hatfield, Gary, 34 n. 11
Hempel, Carl, 45 n. 24, 146 n. 2
Hevley, Bruce, 129 n. 7
Hilbert spaces, 16
Hofweber, Thomas, 67 n. 33
Hume, David, 89 n. 15, 152

identity conditions on what doesn't exist, 59–62, 110–111
implicational opacity, 194
induction, problem of, 152–153
inexpressible physical laws, 26, 41–42, 194
instrumentalism, 34–48. *See also* instrumentalist strategies
instrumentalist strategies, 84, 221
    scientific, 34–39, 41–42, 45–47
    philosophical, 39–43, 45–47, 75 n. 49
isolating a theory, 35 n. 13. *See also* quarantining implications

Jackson, Frank, 58–59

Kant, Immanuel, 7, 9, 89 n. 15, 105 n. 41. *See also* Kant's challenge; ontological independence, Kant's use of
Kant's challenge, 89–91, 90 n. 18, 157–158
Katz, Jerrold J., 101 n. 35
Kreisler, Georg, 18 n. 8

Kim, Jagwon, 153 n. 18
Kitcher, Philip, 146 n. 2
Kripke, Saul, 71, 74 n. 44
Kuhn, Thomas S., 19
Koslow, Arnold, 19n
Krantz, David H., 160, 161

Lakatos, Imre, 37
Landesman, Charles, 54 n. 9, 55–56
Leibniz, Gottfried Wilhelm Freiherr von, 196, 207, 208 n. 57
Lethem, Jonathan, 90 n. 16
Levin, Michael, 45 n. 24
Lewis, David, 62 n. 26, 99 n. 33, 116–117, 148 n. 6
Lipeles, Aaron, 128 n. 4
logic, as a source of ontological commitments, 148
logically isolated theories, virtues of, 152 n. 17

Mach, Ernest, 196, 198, 204 n. 50, 210
Maddy, Penelope, 5 n. 4, n. 5, 29n, 32 n. 8, 38–39, 45–47
Marcus, Ruth. *See* Barcan-Marcus, Ruth
Maris, Humphrey, 214 n. 66
mathematical abstracta
    nonexistence of, 4–5, 109
    ontological dependence of, 103–107, 109, 120–122
    proxying for what is real, 6, 111, 170–171, 172–173, 174, 215, 224
mathematical fictionalism, 48. *See also* Field, Hartry
Maudlin, Tim, 198 n. 39
Maxwell, Grover, 39 n. 20
Millikan, Robert, 213
McConnell, Jeff, 73 n. 42
McGrath, Sarah, 31 n. 6
measurement theory, 160–164
Meinong, A., 56 n. 15
metaphor, necessary condition on, 37–38
metatheoretical reasoning in physics, 24
misappropriated folk psychological notions, 74–75

monitoring, 132–133, 134
    characterization of, 129
Moore, G. E., 113n, 222
Morgenbesser, Sidney, 19n

naturalized epistemology, 149
naturalizing ontology, program of, 5
Ncm. *See* Newtonian cohesive-body mathematics
Nerlich, Graham, 211–212, 214 n. 63
Newton, Isaac, 41, 186 n. 12, 188 n. 18, 191, 196, 197 n. 35, 200, 104–205. *See also* Newton's laws of motion; Newton's law of gravitation.
Newtonian cohesive-body mathematics, 166–171, 181–215, 219–221, 224
    as an ontological program, 187, 187 n. 15 (*See also* privileged frame program, the)
    characterization of, 166
    compared to general relativity, 186
    ontological status of forces within, 182–191 (*See also* forces)
    ontological status of space and time within, 200–201
    quantifier-commitments of, 166–167
Newton's bucket, 204–205, 211
Newton's law of gravitation, 19, 166, 182, 197, 201, 202, 203, 204
Newton's laws of motion, 18–19, 166, 182–185, 188 n. 18, 197, 201, 202, 203, 204
nontriviality requirement, the, 101, 103–104, 106–108, 112, 144. *See also* ontological independence
    characterization of, 100
nontrivially reliable, definition of, 100
Noll, Walter, 169 n.
nominalism, 4, 120, 121–122, 173–174, 181, 224. *See also* deflating ontology
non-literalist strategies, 37–38, 39. *See also* instrumentalist strategies; misappropriated folk psychological notions.

Occam's razor, 85–86, 92, 97, 154 n. 21
Oddie, Graham, 149 n. 8
ontological closure conditions, 150–156
ontological commitment. *See* criterion for what a discourse is committed to; criteria for what exists; folk ontology; logic, as a source of ontological commitments; ontological independence; quantifiers, objectual, ontologically committing by virtue of semantics alone; theory, empirical adequacy of insufficient for commitment; theory, uniqueness of insufficient for ontological commitment; vernacular, indicating ontological commitments within; vernacular "there is," evaluation of ontological commitments of
ontological independence
    inadequacies of coherentist epistemology with respect to, 109, 144–145
    Kant's use of, 105 n. 41
    refinements in characterization of, 113
    reliability requirement on our taking something as, 99–103
ontological nihilism, 5, 223
ontology and ideology, interplay of, 21–22, 25–26
opacity, 70–71, 78. *See also* co-extensionality*

PA. *See* Peano Arithmetic
paraphrase strategy, 38–39
Parsons, Terence, 54 n. 9, 65, 94 n. 25
Peano arithmetic, 17 n. 5, 18, 21, 28, 60, 160, 168, 188, 189
Perrin, Jean, 133
physicalism, 8, 153–154
Plato, 117 n. 7
Platonism, 4. *See also* mathematical abstracta
    Benacerraf's challenge to, 99 n. 33
Plotinus, 8 n. 9

posit
    characterization of, 125–126
    identifications among, 149–158
    thick, characterization of, 129
    thin, characterization of, 128–129
    ultra-thin, characterization of, 127–128
positron, Dirac's prediction of, 178 n. 31
Presburger arithmetic, 18 n. 8
privileged frame program, the, 197–199, 200–209
    characterization of, 197
    global character of, 207 n. 54
    not a matter of simplicity, 199 n. 42
    ontological assumptions of, 201–202
    purported circularity of, 201–203
    underdetermination with respect to, 203–204
properties, ontological dependence of, 107–109
propositions, ontological status of, 111
Ptolemaic astronomy, 34
Purcell, Edward M., 23, 26, 177
Putnam, Hilary, 39 n. 20, 144, 145, 216–217. See also Boyd-Putnam "no miracle" argument; Putnam modals; Quine-Putnam indispensability thesis
Putnam modals, 31–33
Psillos, Stathis, 4n

Q, 18 n. 8
quantifiers
    objectual, 4
        anaphoric role of, 58–59
        definition of, 20 n. 11
        ontologically committing by virtue of semantics alone, 53–55
        substitutional, 20–21
        definition of, 20, n. 11
quantum mechanics, 154–155, 174, 216n, 225n
quarantining implications, 35–39, 46–48, 84
Quine, W. V., 7–9, 20. See also coherentist epistemology; Quine-Putnam indispensability thesis; Quinean rent; regimentation
    his criterion for what a discourse is committed to, 3, 4–5, 6, 10, 49–80, 126–127, 222–223
    appropriate version of, 50
    his five virtues, 128, 129 n. 5, 136, 138, 144
    his physicalism, 8 n. 12
    his term "posit," 125–126
    his triviality thesis, 10, 49, 53, 62–79
    his views on schemas, 20
    on folk ontology, 51n, 114
    on individuation conditions, 59
    on metaphor and ontology, 66n
    on names in the vernacular, 52
    on paraphrase, 38 n. 18
    tensions in his ontological views, 7–8
Quinean rent, 129, 136, 137, 138, 144
    characterization of, 128
Quinean epistemology. See coherentist epistemology
Quine-Putnam indispensability thesis, 4, 29, 47–49, 75–76, 81–82, 200

Rajagopal, K. R., 167 n. 9
recursive axiomatizations, 17
reference*, 70, 82, 149 n. 9. See also co-reference*
    definition of, 61–62
refinement, 131–132, 134, 143
    characterization of, 129
regimentation, 4, 8, 51–52, 63, 80, 114
    normativity of, 51–52
    tension of normativity of with triviality thesis, 53
Reichenbach, Hans, 196
reliability requirement, the, 99–103, 144
    characterization of, 99
Resnik, Michael D., 46, 61 n. 23, 78n, 101 n. 34, 109, 144
robustness, 130–131, 134, 142, 143
    characterization of, 129
Rorty, Richard, 144

Rosen, Gideon, 30 n. 2
Routley, Richard, 54 n. 9, 55–57, 65
rule-following problem, the, 226
Russell, Bertrand, 71, 106, 217–128, 219 n. 73. *See also* Russell's theory of descriptions
Russell's theory of descriptions, 52

Salmon, Wesley C., 217–218
Salmon, Nathan, 65 n. 30, 71–72, 72 n. 39, 74 n. 44
Scheffler, Israel, 50 n. 2
schemas, 19–20, 183–185
Schewe, Phillip F., 214 n. 66
Schiffer, Stephen, 54 n. 9, 93 n. 23, 110–111
scientific realism, success argument for. *See* Boyd-Putnam "no miracle argument"
semantics, ontological neutrality of, 53–62, 78–79, 88–89, 122
separation thesis, the, 5
Shapiro, Stewart, 21 n. 14
simplified models in physics, 23
Sklar, Lawrence, 208 n. 57
Smart, J. J. C., 54 n. 9
Smith, George, 41, 186 n. 11
Sober, Elliot, 29n, 31
space. *See also* space and time
  in contrast with void, 209
  purported causal properties of, 205–207, 209
space and time. *See also* spacetime
  ontological dependence of, 200–201
spacetime
  continuity of, 47
  nonexistence of, 6, 200–201
  substantivalist/relationist debate, 6, 196–212 (*See also* Newton's bucket)
  general form of, 196
space-time points, ontological dependence of, 212–219
standard model, the, 148 n. 5, 151 n. 13
Stein, Ben, 214 n. 66
Stein, Howard, 199 n. 40, 199–200, 202 n. 45
Steiner, Mark, 175–180

Strasberg, M., 135 n. 16
Sussman, Gerald Jay, 22 n. 15

Tarski biconditionals, 16, 18, 26–27, 31–33, 35, 48
Tarski's theory of truth, ontological neutrality of, 53–55, 61–62
Taylor series, 177n
theoretical deductivism, falsity of, 17, 22–25
theory. *See also* approximating theories; black box objection, the; calculational shortcuts; deduction, idealizations of; finite axiomatizability; isolating a theory; metatheoretical reasoning in physics; physicalism; privileged frame program, the; recursive axiomations; schemas; theoretical deductivism, falsity of; theory/observation distinction
  empirical adequacy of insufficient for ontological commitment, 144, 209–211
  operating constitutively or descriptively, 142
  underdetermination of, 216–217
  unifying power of, 146–147
  uniqueness of insufficient for ontological commitment, 146
theory/observation distinction, 39–41
thick epistemic access, 85, 136 n. 17, 136 n. 18, 136–142, 143, 145–146, 148, 149, 151, 170, 171–172, 212–214, 214 n. 66, 223–224. *See also* grounding, refinement, monitoring, robustness
  characterization of, 129
  defeasibility condition on absence of, 138–139
  facticity of, 141, 159
  nontransitivity of across composition, 156, 211–212, 212 n. 61, 212–213
  reasons for epistemic centrality of, 134
  relation to causation, 133–134, 140–141, 101 n. 35

Tomer, Adrian, 132n
top quark, the, 148 n. 5
Toulmin, Stephen, 199–200, 199 n. 42
trivial explanation, the, 100, 102
Truesdell, C., 45, 156 n. 23, 167, 167 n. 8, 167 n. 9, 168 n. 12, 169 n. 13, 169 n.14, 184 n. 7, 191n, 192 n. 27
truth, inflationist views of, 31–33
truth ascriptions
  blind
    definition of, 15–16
    ineliminable, 15–16, 17–26, 28, 29, 34, 44
  explicit, 16
    definition of, 15
    finite statability condition on, 16
    manageable exhibition condition on, 17, 18
    transparency condition on, 16
  joint assertibility of, 27–28
truth predicate
  possible elimination of, 48
  univocality of, 27–28

Urmson, J. O., 75 n. 48
usages of laziness, 63–64, 78

van Fraassen, Bas, 29n, 45 n. 24, 83 n. 3, 84 n. 7
vernacular, indicating ontological commitments within, 103 n. 38
vernacular "exists," 117 n. 7. *See also* folk ontology

as a criterion transcendent term 91 n. 19
contrastive role of, 97, 115–116, 118
ontological neutrality of, 115–117
"really" applied non-redundantly to, 91 n. 19, 116
vernacular "see," 119–120, 158–159
vernacular "there is." *See also* folk ontology
  distinguishing two roles for, 67 n. 22
  evaluation of ontological commitments of, 62–79
  not ambiguous, 78
van Inwagen, Peter, 3 n. 1, 63 n. 26, 65 n. 30, 71, 115
Vinueza, Adam, 92 n. 21, 92 n. 22, 99 n. 32
Vlastos, Gregory, 219 n. 73

Walton, Kendall L., 69n, 73–74, 73 n. 43, 74 n. 46
Weinberg, Steven, 8, 152n
Weiskrantz, L., 132n
Whewell, William, 146 n. 2
Wilson's cloud chamber, 133
Wisdom, Jack, 22 n. 15
Wittgenstein, Ludwig, 88n
Wright, Crispin, 31 n. 5

Yablo, Stephen, 39, 66n, 73 n. 42, 76 n. 50

Zeno's arrow paradox, 217–219